T0271341

High
Temperature
Corrosion

High Temperature Corrosion

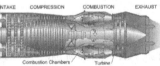

A. S. Khanna
Indian Institute of Technology, Bombay, India

World Scientific

NEW JERSEY · LONDON · SINGAPORE · BEIJING · SHANGHAI · HONG KONG · TAIPEI · CHENNAI · TOKYO

Published by

World Scientific Publishing Co. Pte. Ltd.

5 Toh Tuck Link, Singapore 596224

USA office: 27 Warren Street, Suite 401-402, Hackensack, NJ 07601

UK office: 57 Shelton Street, Covent Garden, London WC2H 9HE

Library of Congress Cataloging-in-Publication Data
Names: Khanna, Anand S.
Title: High temperature corrosion / [edited by Anand S. Khanna].
Description: New Jersey : World Scientific, 2016. | Includes bibliographical references and index.
Identifiers: LCCN 2015031963 | ISBN 9789814675222
Subjects: LCSH: Corrosion and anti-corrosives. | Materials at high temperatures.
Classification: LCC TA462 .H5236 2016 | DDC 620.1/1223--dc23
LC record available at http://lccn.loc.gov/2015031963

British Library Cataloguing-in-Publication Data
A catalogue record for this book is available from the British Library.

Desk Editor: Kalpana Bharanikumar

Typeset by Stallion Press
Email: enquiries@stallionpress.com

Printed in Singapore

Foreword

High Temperature Corrosion is a serious problem for many industries. Fossil-fueled power plants, refineries, petrochemical plants, chemical processing plants, operating at temperatures ranging from 400°C to 1100°C, have severe corrosion problems. Fireside corrosion in coal-based power plants, steamside corrosion, sulfidation and carburization problems in refineries and petrochemical plants, and direct chemical attack in the chemical process industries are a few problems which still need solution. Solution to high temperature corrosion problem can be tackled in two ways, first by properly selecting the materials and second by controlling the environment. I think the better solution is proper selection of materials. Development in metallurgy over the years has given several new and advanced materials, starting with specialized steels, stainless steels, superalloys, ceramics and composites. However, selection for specific application needs knowledge of metallurgy, materials science and corrosion. Degradation of materials at high temperatures occur due to both loss in mechanical properties as well as degradation due to corrosion, which can be oxidation, sulfidation or hot corrosion. Hence selection of materials must take into account the synergistic effect of mechanical properties and corrosion. This has been very nicely covered in Chapter 2 from a beginner's prospective and can guide in proper selection of materials at high temperatures. This information coupled with basic knowledge of high temperature corrosion, given in Chapter 1, in terms of stability and kinetics for various advanced materials, can ease the selection process. Chapter 3 provides the basis of the development of advanced materials. Chapters 4–6 are the backbone of this book that present the industrial scenario of degradation of materials in three important sectors — chemical and petrochemicals industry, power plants and aerospace industry. It is well known that even a highly well-developed material still needs protection at the surface. Properties such as hardness, corrosion protection cannot be achieved by bulk alloy development. Hence, Chapter 7 is devoted to various kinds of coatings used to protect materials at high temperatures. To understand the mechanism of oxidation process, it was planned to introduce a chapter on analytical tools to analyze corrosion products. But instead, a specific versatile technique has been included in Chapter 8 which can give very useful information about oxides scales and corrosion layers formed during high temperature exposure. No book can be complete without talking about nanotechnology. The next chapter, that is Chapter 9 focuses on a very important aspect of high temperature corrosion that is active element effect. The chapter very nicely summarizes the present understanding of active element addition and on the stability of oxide layers. A specific example of the use of nanoparticles in the development of high temperature resistant surface coatings has been covered. The last chapter, that is Chapter 10 includes information on reactive element effects.

In short, the book provides fundamental knowledge about high temperature corrosion, development of advanced materials, interaction of corrosion and mechanical properties, corrosion related problems in three important industries and topics such as coatings and nanotechnology. The book can be taken as an advanced information for this topic and can be used as a reference book as well.

It is my pleasant duty to thank all my authors who have contributed the various chapters. Prof. Ukai of Hokkaido University, Japan, Dr. Kangas from Sandvik India, Prof. Raman Singh and his colleague Mahesh from Monash University, Dr. Joe Quaddakers and his Colleague Dimitri from Forschungs Zentrum Juelich, Germany, Dr. Subrato Mukharjee and my Student Nirav Jamanapara from

IPR, Gandhinagar, India and finally to my friend Dr. Vinod Agarwala, retired scientist, US Navy, for helping me on the chapter on corrosion problems in aerospace industry.

I sincerely hope this book will be of immense benefit to new readers and will act as a source book for personnel working in power plants, refineries and aerospace industry.

Prof. A. S. Khanna,
Humboldt Fellow,
Fellow of Royal Norwegian Science and Technology,
Fellow Japan Key Centre,
Fellow ASM International and NACE International,
Chairman SSPC India,
Department of Metallurgical Engineering & Materials Science,
IIT Bombay, India.

Contents

**Chapter 4. High Temperature Corrosion
Problems in Refineries, Chemical
Process Industries and
Petrochemical Plants   87**

Pasi Kangas

**Chapter 5. High Temperature Corrosion
Problems in Coal-based Thermal
Power Plants   99**

A.S. Khanna

**Chapter 6. High Temperature Corrosion
            Problems in Aircrafts   129**

A.S. Khanna and Vinod S. Agarwala

**Chapter 7. Coatings for High Temperature
            Applications   161**

N. I. Jamnapara and S. Mukherjee

Chapter 8. Advanced Analytical Tools to Understand High Temperature Materials Degradation — Ion Beam Characterization of Aerospace Materials 201

Barbara Shollock and David McPhall

**Chapter 9.   Role of Nanotechnology in Combating
                 High Temperature Corrosion   219**

R.K. Singh Raman, B.V. Mahesh and Prabhakar Singh

**Chapter 10. Reactive Element Additions
                 in High Temperature Alloys
                 and Coatings   245**

D. Naumenko and W.J. Quadakkers

Overview

1. **Fundamentals of High Temperature Oxidation/ Corrosion**

   Prof. A. S. Khanna, IIT, Mumbai, India

   Criteria of metal oxidation, sulfidation and hot corrosion at high temperatures, kinetics of oxidation, sulphidation and hot corrosion, modeling to assess the life and durability of materials at high temperatures in different environments.

2. **Degradation of Mechanical Properties of Materials at High Temperatures in Corrosive Environments**

   A. S. Khanna, IIT Bombay, India

   Degradation of mechanical properties such as tensile strength, fatigue and creep at high temperature with and without the presence of corrosive gases such as air, oxygen etc. Models to understand the synergistic effect of temperature and environment on mechanical properties.

3. **Materials Development Aiming at High Temperature Strengthening — Steels, Superalloys to ODS Alloys**

   S. Ukai, Hokkaido University, Japan

   Fundamental metallurgical principles to develop high temperature materials, role of microstructure, alloying, strengthening mechanism and method of fabrication on the high temperature properties of materials.

4. **High Temperature Corrosion Problems in Refineries, Chemical Process Industries and Petrochemical Plants**
Pasi Kangas, Sandvik Materials Technology, Pune, India

   Material failure leading to accidents and plant shut down in various chemical process industries, refineries and petrochemical plants, use of better materials, proper control of environments, especially sulfur bearing gases and development of models to select better materials.

5. **High Temperature Corrosion Problems in Coal-Based Thermal Power Plants**
A. S. Khanna, IIT, Mumbai, India

   Complete description of various kinds of power plants, fossil fueled, fluidized bed, nuclear reactors, fireside problems in fossil fueled power plants, fuel/clad problems in nuclear reactors, steam side corrosion problems, new materials for super critical power plants.

6. **High Temperature Corrosion Problems in Aircrafts**
A. S. Khanna and Vinod S. Agrawal

   Aerospace industry uses one of the most corrosion resistant materials. Description of various degradation mechanisms in aircraft gas turbines, combustion chamber and use of advanced materials.

7. **Coatings for High Temperature Applications**
N. I. Jamnapara and Subrato Mukherjee, Institute of Plasma Research, Gujarat, India

   Starting with surface modification methods, classification of various high temperature coatings, based on functional application, techniques used, detailed methodology with pros and cons of the coating, examples of a few specific examples where role of coating has been emphasized.

8. **Advanced Analytical Tools to Understand High Temperature Materials Degradation — Ion Beam Characterization of Aerospace Materials**
Barbara Shollock, University of Warwick, UK and David S. McPhail, Imperial College of London, UK

Starting from basics of optical and electron optical techniques, a coverage of advanced techniques such as AES/ESCA, SIMS, focused Ion beams, AFM etc.

Addresses of Corresponding Authors

1. A. S. Khanna,
   Department of Metallurgical Engineering & Materials Science,
   Indian Institute of Technology, Mumbai – 400076, India.
   anandkh52@ggamil.com.

2. S. Ukai,
   Laboratory of Advanced High Temperature Materials,
   Research Group of Energy Materials, Division of Materials
   Science and Engineering,
   Hokkaido University, Japan.
   s-ukai@eng.hokudai.ac.jp.

3. Pasi Kangas,
   Sandvik Materials Technology
   Mumbai Pune Road, Dapodi, Pune 411012, India.
   passi.kangas@sandip.com

4. S. Mukherjee,
   Institute of Plasma Research, Gandhinagar, Gujarat, India.
   mukherji@ipr.res.in

5. David S. McPhall,
   Department of materials
   Imperial College London
   Room No. 1.02, Royal School of Mines Building
   South Kensington Campus
   London – SW7 2AZ, UK.
   d.mcphall@imperial.ac.uk

6. Raman Singh,
   Department of Mechanical & Aerospace Engineering
   Department of Chemical Engineering
   Bldg 31, Monash University — Clayton Campus (Melbourne)
   Vic. 3800, Australia.
   raman.singh@monash.edu
7. D. Naumenko
   Institute for Energy and Climate Research (IEK-2)
   Forschungszentrum Jülich GmbH 52425
   Jülich, Germany.
   D.naumenko@fz-juelich.de

Biography

Dr. A. S. Khanna is Professor at The Indian Institute of Technology, Bombay, India with responsibility for teaching, research and consultancy in the fields of corrosion, coatings, surface engineering and corrosion management.

Prior to joining IIT Bombay, Prof. Khanna worked in the department of Atomic Energy with a focus on high temperature corrosion problems on power plant materials. He visited several international labs/universities/institutions, including Forschungszentrum, Juelich, Oslo University, University de Provence, Marseille France and IHI Heavy Industry, Japan.

His professional interests focus on coatings, industrial corrosion prevention, surface engineering, high temperature materials, high temperature coatings, laser surface modifications. He has already guided 22 PhD's, out of which more than 14 are on high temperature corrosion, high temperature coatings and laser alloying. His current projects include development of smart coatings and nanotechnology for enhancing organic coating performance. He is Consultant/ Advisor to many industries, including oil and gas, refineries, power plants and petrochemical plants. He has written two books, one is on High Temperature Corrosion, published in 2002 by ASM

International, Ohio, USA. His second book is on High Performance Coatings, published by Woodhead publication UK in 2008. He has also edited four conference proceedings. In addition, he serves as Chairman for SSPC India, and is Fellow of NACE International and Fellow of ASM International.

Chapter 1

Fundamentals of High Temperature Oxidation/Corrosion

A.S. Khanna

Department of Metallurgical Engineering and Materials Science
Indian Institute of Technology
Mumbai 400076, India

Material degradation at high temperatures takes place due to loss in mechanical properties with increase in temperature as well as due to the chemical interaction of metal with the environment. This chemical interaction is further sub-divided into oxidation, sulfidation, and hot corrosion. While oxidation leads to the formation of oxide, which can be deleterious if the oxide is fast growing and spalls extensively, however, if the scale formed is adherent, thin and slow growing, it provides protection to the base metal or alloy. Sulfidation is a much severe degradation process and several times faster than the oxidation. In many industrial environments, it is a mixed gas environment, leading to oxidation and sulfidation simultaneously. Hot corrosion is another degradation mechanism which is even more severe than the oxidation and the sulfidation. Here, oxidation/sulfidation occurs in the presence of a molten salt on the surface of the substrate. Related issues, such as role of defect structure, active element effect and stress generation, during oxide growth process, have also been discussed. Finally, a guide to material selection for high temperature application is presented.

1. Introduction

Corrosion eats away several billions of our hard earned money in replacement, repair and maintenance of several components, parts or the whole equipment, due to leakage, catastrophic accidents or due to plant shut down. One of the main reasons of corrosion of metals and alloys is the reaction with environment around it, which may be natural atmosphere, or the liquid or gaseous environment around the metallic component. Many corrosion failures occur at room temperature for which some electrolyte is necessary. However, at high temperatures, the corrosion occurs due to direct reaction of metal with its environment. For example, steel exposed to oxygen at room temperature does not cause any reaction of oxygen with metal, however if the same steel is exposed to oxygen at $600°$C, the steel reacts with oxygen, forming various iron oxides on the surface. Thus, the main criteria of corrosion of a metal under the gaseous environment at high temperature is its tendency to form an oxide or other products, for which the free energy of the formation must be large negative. For example, for a reaction,

$$M(s) + O_2(g) = MO_2(s). \tag{1}$$

$\Delta G°$ should be large negative. $\Delta G°$ is the standard free energy for the formation of the oxide MO_2. The free energy can be related to partial pressure of oxygen using a standard equilibrium condition by

$$\Delta G° = RT \ln pO_2(g), \tag{2}$$

where $pO_2(g)$ is the partial pressure of oxygen at a temperature T and R is the gas constant. This is a very useful equation from which we can get the dissociation pressure of oxygen at any temperature if we plot $\Delta G°$ versus temperature. Also, we can get the relative stability of various oxides of several metals in the periodic table. Further, using this table, one can read the dissociation pressure of various metals at different temperatures and can also find out the minimum partial pressure, required for a metal to oxidize at various temperatures from a well-designed nomo-graphic scale around the $\Delta G°$ versus T plot. This diagram is well known as Ellingham diagram or Richardson

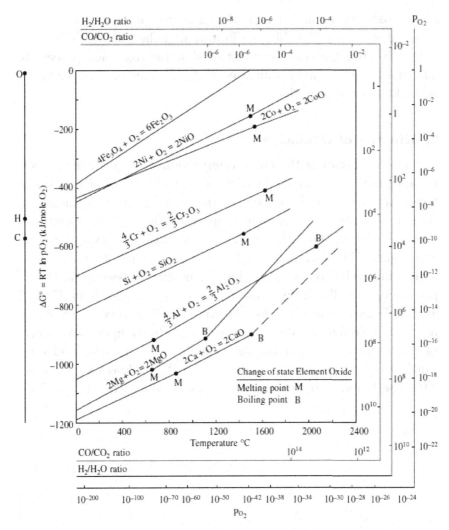

Fig. 1.   Ellingham/Richardson diagram.[1]

diagram and is shown in Fig. 1. There are three nomo-graphic scales, where one can read directly, the $pO_2(g)$, ratio of $p(H_2/H_2O)$ and $p(CO/CO_2)$ which are related to $pO_2(g)$. Details of this diagram can be seen in Ref. 1

Ellingham diagram, therefore, defines the oxidation process due to the basic thermodynamic criterion. It simply tells whether a metal can be oxidized or not at a particular temperature and at a given pressure of oxygen. The biggest limitation of the Ellingham diagram is that it cannot predict how fast or slow the oxidation process is. Therefore, another important aspect of high temperature oxidation is the kinetics of oxidation.

2. Kinetics of Oxidation

Oxidation kinetics is the engineering requirement of high temperature oxidation. Every engineer requires the lifetime of a metal in terms of its oxidation resistance. Hence, it is very important to predict the life of a component operating at high temperatures. Kinetic behavior basically means the variation of oxidation rate with time. This variation can be logarithmic, parabolic or linear with time, giving rise to three important kinetic laws: logarithmic, parabolic, and linear kinetics, respectively, as shown in Figs. 2 and 3. Logarithmic oxidation kinetics predicts very quick initial reaction, followed by almost no reaction. Based upon how slowly the rate subsides after initial fast reaction, the logarithmic kinetics can have direct or inverse behavior of scale thickness versus time. This law is followed by almost all metals when they are oxidized

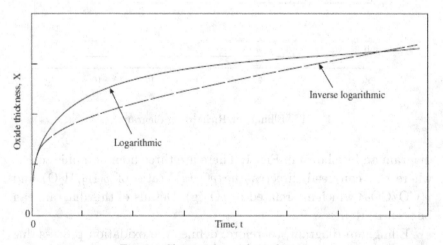

Fig. 2.   Showing logarithmic kinetics.

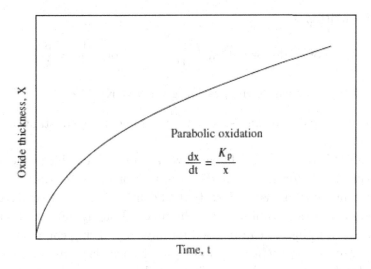

Fig. 3. Showing parabolic kinetics.

at low temperatures and low pressure or for noble metals at high temperatures. The kinetic equation of the two laws is written as

$$X = k \log t \quad \text{Direct Logarithmic law,} \tag{3}$$

$$\frac{1}{X} = K + \log t \quad \text{Inverse Logarithmic Law.} \tag{4}$$

The most important law which is usually followed by important high temperature metals and alloys is parabolic law. This law predicts that the rate of oxide layer formation is inversely related to time, which means that as the time increases, the rate of scale formation is continuously decreasing. As a matter of fact, every engineer would like to choose a metal whose oxidation rate decreases with time. Basic reason behind this law is that here oxidation is under diffusion control. It is assumed that the oxidation occurs either by the diffusion of metal ions from metal substrate towards oxide gas interface to react with oxygen, or oxygen ions diffuse through the oxide layer, reach the metal/oxide interface and react with metal. With time, as the scale growth takes place, diffusing species requires longer time to diffuse thicker oxide layer. The parabolic kinetics is usually written as

$$\text{Rate of oxide formation} \left(\frac{\mathrm{d}x}{\mathrm{d}t}\right) \infty \frac{1}{x} \quad \text{or} \quad \frac{\mathrm{d}x}{\mathrm{d}t} = \frac{K}{x}, \tag{5}$$

$$\text{which on integration gives } x = Kp \; t^{-\frac{1}{2}}, \tag{6}$$

where x is scale thickness, Kp is parabolic rate constant and t is time.

The third law is the linear law, according to which, the rate of oxide scale formation is directly proportional to time, which means the reaction is so fast that the metal reacts with oxygen, as soon it comes in contact with the metal. Usually, no metal which follows linear kinetics can be used for any engineering component. In high temperature oxidation, indication of linear law basically means some kind of catastrophic reaction, resulting due to the cracking of oxide scale or other failures such as scale delamination or spallation of oxide. For example, when a metal, which is showing parabolic behavior for a very long time, suddenly shows linear behavior, it indicates some defect in the protective oxide layer, cracking or delamination.

3. Isothermal Versus Cyclic Oxidation

One of the very important criteria to select material for high temperature oxidation resistance is the stability of the oxide layer formed. It is expected that during thermal cycling, the oxide remains intact on the surface without any cracking or spalling. This criterion can be met only if the metal/alloy passes the cyclic oxidation test.

Usually, the oxidation tests are carried out at a fixed temperature and weight gain is measured as a function of time from which kinetics are predicted. This situation of testing is called isothermal condition and there is very little chance metals show cracking or spalling during isothermal oxidation. In actual practice, many high temperature components are heated for certain time, followed by bringing them to room temperature, heating again and followed by cooling and heating several times. A set of heating and cooling is called one cycle. Depending upon the design of a material and its requirements, type and number of cycles are selected and tests are carried out for such number of cycles. For an excellent oxide for high temperature application, the oxide layer must remain intact throughout the test. The reason why oxide scale spalls during thermal cycling is due to release of thermal stresses generated due to sudden cooling from high temperature. Following equation gives the stress generated when an oxidizing sample is cooled from a high temperature $T2$ to a lower temperature $T1$.[1]

$$\sigma \text{ oxide(compressive)} = \frac{Eo\Delta T(\alpha o - \alpha m)}{1 + 2\left(\dfrac{E_o}{E_m}\right)\left(\dfrac{t_o}{t_m}\right)}, \tag{7}$$

where σ is the compressive stress due to t_o thermal cycling, E_o and E_m are, respectively, the young's modulus of oxide and metal with t_o and t_m as respective thicknesses and α_o and α_m are the coefficients of thermal expansion of oxide and metal, respectively, and ΔT is the temperature drop from $T2$ to $T1$.

One of the precautions to avoid spalling or cracking during thermal cooling is to carry out slow cooling, which does not generate enough stresses and thus oxide remains intact. Another method to make spalling resistant alloy is by addition of small concentration of active elements such as yttrium, cerium or lanthanum. These elements make a strong oxide to metal bond by one or more of the various theories, pegging, enhancing plasticity of scale, etc.[1]

4. Oxidation of Pure Metals

A pure metal can form a single oxide or multiple oxides when oxidized at high temperatures. Ellingham diagram guides the partial pressure of oxygen required for a particular type of oxide formed on a metal. The oxidation process is relatively simple when a single oxide layer is formed. After knowing the type of oxide, it is possible to estimate the diffusing species, metal ion or oxygen, which drives the oxide formation. Let us take the example of the oxidation of a pure metal like nickel. Ni, on oxidation forms a single oxide NiO, which is a p-type oxide, thus the Ni^{++} ions are main diffusing species from the nickel substrate to oxide/gas interface and react with oxygen. Usually, a single compact NiO scale is formed when nickel is very pure. Commercial purity nickel has impurity of carbon. This results in a two way diffusion of Ni^{++} ions and O^{--} ions, resulting in a duplex oxide layer, the inner one is quite adherent and the outer a bit porous.[1]

A slightly complex example of oxidation is the oxidation of iron, which forms three oxides, FeO, Fe_3O_4 and Fe_2O_3, when oxidized at temperature above 570°C. At temperature below 570°C, it forms two oxides Fe_3O_4 and Fe_2O_3 as per the Fe–O phase diagram shown in Fig. 4. Schematic of the oxide formed above 570°C is shown in Fig. 5, which depicts the diffusion of Fe^{++} ions at the interface Fe–FeO and FeO–Fe_3O_4 as both these oxides are p-type oxides. Fe_2O_3 is amphoteric oxide, grows by diffusion of both Fe^{+++} ions outward and O^- ions inwards. The corresponding reactions at various interfaces are also given. Thus, it is very clear that the characteristic diffusing species remains intact even when multiple oxides are formed on a metal.

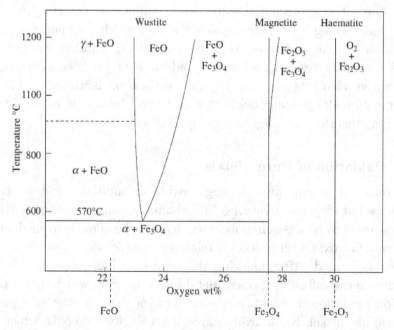

Fig. 4.   Fe–O phase diagram.[1]

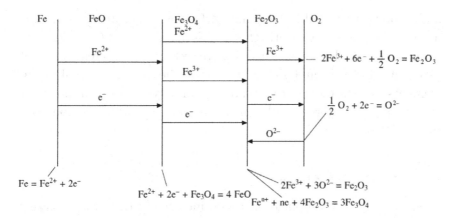

Fig. 5.   Schematic of oxidation of pure iron at temperature above $570°C$.[1]

Another important finding from iron oxidation above $570°C$ is that main species responsible for iron oxidation is FeO whose thickness is 95%, compared to 4% of Fe_3O_4 and 1% Fe_2O_3 layer. Further, the FeO scale is rather fluffy and highly porous and very defective and grows very fast. Thus, if the oxidation of iron has to be reduced, the only way is to reduce or avoid the formation of FeO. That is why the best way to use steel at high temperatures is to alloy it with Cr. There are important alloys of Fe and Cr, ranging from 1% to 9% of Cr. Some of the common Fe–Cr alloys, used for moderate temperature (250–540$°C$) applications are 21/4 Cr–1Mo and 9Cr–1Mo steels. These two alloys are important material for heat exchanger tubes, economizer and superheater tubes, respectively.

Oxidation protection of these alloys is due to the formation of a spinel oxide of formula $FeO \cdot Fe_xCr_{2-x}O_3$, which initially forms an iron rich mixed oxide with chromium and then changes to chromium rich spinel at higher levels of Cr in the alloy. And when the Cr concentration crosses 11.5%, it forms a selective chromium oxide layer which protects the steel from further oxidation. At this concentration, the steel turns to stainless.

Other important metals, considered for high temperature oxidation are titanium and zirconium. Both have a tendency to dissolve oxygen in addition to stable oxide formation. Both Ti and Zr form stable protective oxides TiO_2 and ZrO_2, which protect the metals at high temperatures. Both the metals have α and β phases, the former has a high tendency to dissolve oxygen which creates lots of problems. Both Ti and Zr are therefore alloyed with β stabilizing elements which reduce the tendency of oxygen dissolution to a greater extent. More details on Ti and Zr oxidation can be seen in Refs. 2–4.

5. Oxidation of Alloys

Consider how the complication would increase further, when instead of a pure metal, the substrate consists of more than one metallic element. In principle, as per Ellingham diagram, all the metallic elements present can oxidize and form their respective oxides, based upon the oxygen partial pressure, available at that temperature. This is called transient oxidation. However, when several oxides are present, again as per Ellingham diagram, the weaker oxide would vanish and only stable oxide will remain. However, this process is not very simple and it depends on several factors such as concentration of alloying elements, the stability of their oxides, temperature and their diffusion coefficients. Thus, in a multielement alloy, there can be several processes which need to be considered before a final oxide layer is decided during oxidation process.

Two very important terminologies during oxidation of alloys are selective oxidation and internal oxidation. It is very important to understand the behavior of alloys and the conditions under which they form selective or internal oxidation. Internal oxidation is a process when an oxide is formed within the matrix of the base alloy, just below the external oxide. This possibility occurs when the concentration of alloying element is very less and takes more time to diffuse to oxide/gas interface. Under such condition, the oxygen diffuses faster from environment through oxide and reaches oxide metal interface, further diffuses into metal and forms oxide wherever it meets the low concentration diffusing metal ion, resulting in precipitation of the oxide within the substrate matrix. This condition of low concentration and low diffusion coefficient of alloying element, compared to large diffusion of oxygen and higher oxygen solubility, is shown in the following equation:

$$N_m * D_m \ll N_s * D_o, \tag{8}$$

where N_m is the concentration of metal ions in the alloy and D_m is its diffusion coefficient in the metal while N_s and D_o are respectively the solubility and diffusion coefficient of oxygen.

When the concentration of the oxide forming element reaches beyond a critical concentration, termed as selective oxidation concentration, then this element diffuses faster towards the metal/gas interface and immediately forms a selective oxide scale of that element which protects it from further oxidation. This is basically the fundamental of protection of alloys at high temperatures. Presence of 11.5% Cr in ferritic stainless helps in protecting it from further oxidation by forming a protective chromium oxide layer. The condition of selective oxidation is given by equation,

$$N_m * D_m \gg N_s * D_o. \tag{9}$$

The most important high temperature alloys are stainless steels and superalloys. Both these alloys are based upon the formation of a protective oxide layer, chromia or alumina, the formation of which protects them from further oxidation.

5.1 *Oxidation of Stainless Steels*

The name, stainless steel, comes from a unique combination of Fe and Cr. When Cr concentration in a Fe–Cr alloy reaches 11.5%, the alloy does not stain even when exposed to wet or moist environment. At this concentration, the Fe–Cr alloy forms a very thin passive layer of chromium oxide, which is responsible for making this alloy stainless. The stainless steel alloy, so formed is called Ferritic Stainless Steel, whose structure is same as that of the steel. There are various ferritic stainless steels, as listed in Table 1 with their chemical compositions and mechanical properties. Many of these stainless steels are used for high temperature applications, for example, Alloy 430 is considered to be an important high temperature stainless steel.

A.S. Khanna

Table 1:   List of various ferritic stainless steels with their mechanical properties and high temperature capability.

Grade	C	Mn	Si	Cr	Mo	P	S	Other elements	Tensile strength MPa
405	0.08	1.0	1.0	11.5–14.5	—	0.04	0.03	0.1–0.3 Al	415
409	0.08	1.0	1.0	10.5–11.75	—	0.045	0.045	(6xC) Ti min	380
429	0.12	1.0	1.0	14.0–16.0	—	0.04	0.03		450
430	0.12	1.0	1.0	16.0–18.0	—	0.04	0.03		415
446	0.20	1.5	1.0	23.0–27.0	—	0.04	0.03	0.25 N	480

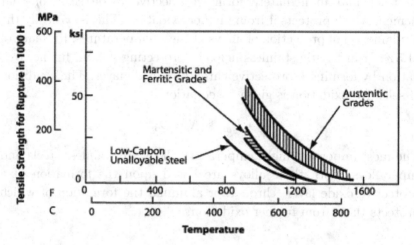

Fig. 6.   Comparison of tensile strength of steels and various stainless steels.[5]

Though the ferritic stainless steels are good for several moderate temperature applications, their use is restricted from their high temperature capability. The temperature limit of ferritic stainless steels is usually 600°C, after which they lose their mechanical strength very rapidly as shown in Fig. 6 in the comparison of mechanical strength of various other steels and stainless steels. Thus for temperatures above 600°C, usually austenitic stainless steels are used. An austenitic phase is a high temperature phase which can be made stable at room temperatures by the addition of austenitic stabilizing elements such as Ni or Mn.

The most common, all-purpose austenitic stainless steel is called 18/8 stainless steel which is represented by UNS nomenclature as AISI Type 304. This stainless steel is formed by the addition of 8% of Ni and 18% of Cr, which results in the formation of a selective chromium oxide scale which protects it from further oxidation. For longer high temperature oxidation durability, just 18% of Cr is not adequate. Higher concentration of Cr is good to create a long lasting protective oxide scale. A list of some of the important austenitic stainless steels of 300 series is given along with their mechanical properties in Table 2.

5.2 *Oxidation of Superalloys*

Next class of high temperature alloys are superalloys. Superalloys, from their basic definition, are alloys which sustain high temperature strength and possess strong oxidation resistance at high temperatures. Unlike austenitic stainless steels, which lose their strength above 750°C, superalloys retain strength till 1050°C, as is evident from Fig. 7. This retention of strength at high temperatures is due to the presence of several substitutional solid solutioning

Table 2:   A list of austenitic stainless steels with their mechanical properties and temperature capability.[6]

AISI grade	C max.	Si max.	Mn max.	Cr	Ni	Mo	Ti	Nb	Tensile strength (MPa)
301	0.15	1.00	2.00	16–18	6–8				515
302	0.15	1.00	2.00	17–19	8–10				515
304	0.08	1.00	2.00	17.5–20	8–10.5				515
310	0.25	1.50	2.00	24–26	19–22				515
316	0.08	1.00	2.00	16–18	10–14	2.0–3.0			515
321	0.08	1.00	2.00	17–19	9–12		5 x %C min.		515
347	0.08	1.00	2.00	17–19	9–13			10 x %C min.	515

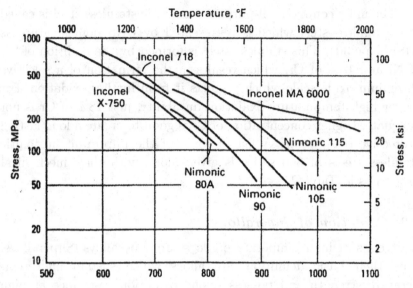

Fig. 7.   Tensile strength as a function of temperature (°C) for various superalloys.[7]

strengthening elements such as Mo, Nb, Ti, ad W along with strong intermetallic precipitates such as Ni_3Al, Ni_3Ti and/or Ni_3Nb. These intermetallic precipitates help to retain the strength of superalloys by providing strong obstacles to the dislocation movement.

Superalloys are basically Ni base or Fe–Ni base or Co base alloys with a suitable concentration of Cr and Al to form either a protective chromia layer or alumina layer, which protects it from further oxidation. Based upon these, the superalloys are called chromia formers whose temperature limit is 900°C, as chromia gets destabilized above 900°C, and alumina forming alloys whose temperature capability is up to 1150°C. The chromia forming alloys have Cr concentration ranging from 10% to 20% or more, while alumina forming alloys must have at least 10% Cr and 5% Al. Higher Cr containing superalloys form alumina layer, even at slightly lesser Al concentration. Gathering of Cr is must to form alumina layer because of two reasons: (i) it helps to avoid any internal aluminum oxide formation due to gathering of oxygen and (ii) it also helps in forming a more stable α-alumina.[1] Superalloys have mostly austenitic structure. Some of the important superalloys with their composition, mechanical properties and whether chromia or alumina forming type are given in Table 3.

6. Sulfidation

Second most serious degradation mechanism at high temperatures is the sulfidation. Here, the environment is either pure sulfur vapor, reducing gasses such as hydrogen sulfide or oxidizing and sulpfidizing environment, sulfur dioxide or trioxide. While pure sulfide formation takes place in case of pure sulfur vapors or H_2S, in case of SO_2/SO_3, the degradation can lead to the formation of oxide or/and sulfide, depending on the oxygen partial pressure. Sulfidation is a severe problem in many industrial environments such as refineries, petrochemical plants, desulfurization plants, etc. Sulfidation is several order of magnitude faster than the oxidation, as shown in Table 4, from the parabolic rates of some metal sulfides and oxides. This is mainly because of the fact that sulfides are more defective, they form low melting eutectics, they do not form any selective sulfide as it happens in the case of selective oxide formation such as chromia/alumina, which once formed, protects the base alloy from further oxidation.

One of the most important aspects of sulfidation is to select materials for sulfidation environments. The steels are very reactive to sulfur vapor or H_2S. Alloying with chromium helps to reduce sulfidation. The higher the chromium level, the lower the extent of sulfidation. The selection of different alloys, steels or stainless steels is based upon the temperature and the concentration H_2S vapor in the system. Figure 8 depicts the iso-corrosion diagrams [9] for steel and stainless steels. Proper steel composition can be chosen, by fixing the temperature and H_2S partial pressure, to select a desired steel or stainless for a certain corrosion rate.

Table 3: A list of important superalloys with their composition, mechanical properties and type.[8]

Alloy	Type	Cr	Co	Mo	W	Ta	Nb	Al	Ti	Fe	C	B	Zr	Re	Hf	Y	Tensile strength
Inconel718	Wrought	18.6	—	3.1	—	—	5.0	0.40	0.9	18.5	0.04	—	—	—	—	—	—
Inconel 600	Wrought	15.8	—	—	—	—	—	—	—	7.2	0.04	—	—	—	—	—	—
Nimonic 80	Wrought	19.5	1.1	—	—	—	—	1.3	2.5	—	—	0.06	—	—	—	—	—
Rene 41	Wrought	19	11	10	—	—	—	1.5	3.1	—	0.09	0.05	—	—	—	—	—
Udimet 500	Wrought	18	18.5	4	—	—	—	2.9	2.9	—	0.08	0.006	0.05	—	—	—	—
Hastelloy X	Wrought	22	1.5	9	6	—	—	—	—	18.5	0.1	—	—	—	—	—	—
Waspaloy	Wrought	19.5	13.5	4.3	—	—	—	1.3	3	—	0.08	0.006	0.06	—	—	—	—
Astroloy	PM	14.9	17.2	5.1	—	—	—	4	3.5	—	0.03	—	0.04	—	—	—	—
MA 758	ODS	30	—	—	0.5	—	—	0.3	—	—	0.04	—	—	—	—	—	—
TMS 63	Single Crystal	6.9	—	7.5	—	8.4	—	5.8	—	—	—	—	—	—	—	—	—

Table 4: Parabolic oxidation rate versus sulfidation rate for a few important metals.[1]

Environment	Parabolic rate constant (Kp) gm^{2}cm^{-4}s^{-1}			
	Ni	Co	Fe	Cr
Oxidation	9.1×10^{-11} $1000°C$	1.6×10^{-9} $950°C$	5.5×10^{-8} $800°C$	4.5×10^{-12} $1000°C$
Sulfidation	8.5×10^{-4} $650°C$	6.7×10^{-6} $720°C$	8.1×10^{-6} $800°C$	8.1×10^{-7} $1000°C$

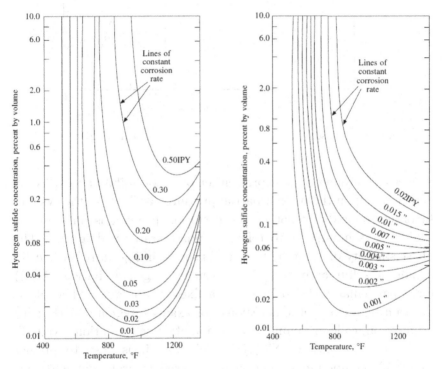

Fig. 8. Mcnormy diagrams showing iso-corrosion curves for steel and stainless steels at various partial pressures of H_2S.[9]

Many industrial environments do not have pure sulfur vapors or reducing gases such as H_2S. SO_2 is a very common gas present in many industrial systems. For example, parts per million (ppm) level of "S" impurity in coal can create SO_2/SO_3 gases in the fire side.

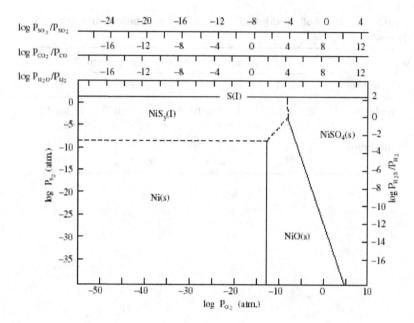

Fig. 9.   Phase stability diagram for Ni–O–S system at 627°C.[10]

Such an environment can form either oxide or sulfide on exposure, depending upon the oxygen partial pressure, available in the system. In order to predict whether the corrosion product is an oxide or sulfide, it is important to refer the phase stability diagrams. The phase stability diagram of Ni metal is shown in Fig. 9. It is basically a plot of sulfur activity $(pS_2(g))$ versus oxygen activity $(pO_2(g))$ for a particular metal or alloy at a fixed temperature. The phase diagram highlights various regions, for example region "Ni(s)" with no reaction, called immune area, region of oxide formation "NiO(s)" or sulpfide formation "NiS$_y$(I)". This way it is easy to predict the stable compound formation in a mixed environment and thus understand the effect of complex environment on corrosion of base metal.

In many industrial environments, sulfur activity is not available in the form of $(pS_2(g))$, but in the form of sulfur dioxide. Now in order to predict what could be the corrosion product at a particular SO_2 environment, one has to draw a trace of SO_2 from the relationship of SO_2 partial pressure with sulfur or oxygen activity as given in the following equation:

$$S_2(g) + 2O_2(g) = 2SO_2(g), \tag{10}$$

$$\Delta G^\circ = RT \ln pSO_2(g)/(pO_2(g))^2 pS_2(g), \tag{11}$$

$$\ln pO_2(g) = \frac{\Delta G^\circ}{RT} + 2 \ln pSO_2(g) - \ln pS_2(g). \tag{12}$$

Thus, isobaric lines for $SO_2(g)$ can be drawn, each with a slope of 2 as shown in Fig. 10 for the phase stability diagram of Cr–O–S system. It can be seen that 10^{-2} atm., (point "A" in Fig. 10) pressure of SO_2 lies in Cr_2O_3 region, confirming that at this pressure SO_2 will oxidize and form stable Cr_2O_3, rather than sulfidize chromium. For other partial pressures of SO_2, other stable products such as chromium sulfides and mixed supfide and oxide can form.

In the same way, phase stability diagram of multicomponent alloys such as stainless steel Fe–Cr–Ni alloy can be constructed. This

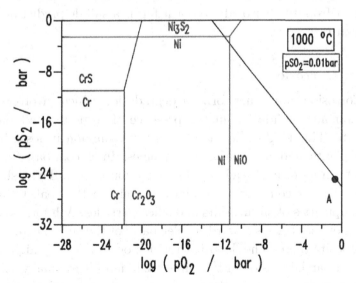

Fig. 10. Phase stability diagram of Cr at 1000°C in SO_2 partial pressure of 0.01 bar.[1]

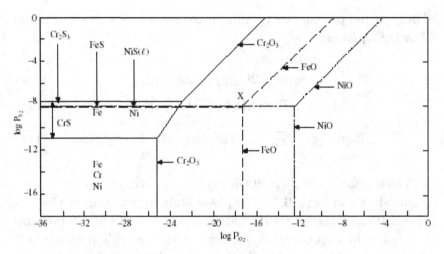

Fig. 11.   Phase stability diagram for stainless steels (Fe–Ni–Cr alloy).[11]

is shown in Fig. 11 for the corrosion at 870°C in mixed environment of oxygen and sulfur bearing gases. From this diagram, it can be predicted at what concentration of oxygen or sulfur activity, the stainless steel will form a stable oxide such as chromium oxide to protect it from further oxidation or it forms a sulfide predicting high corrosion rate.

7. Hot Corrosion

Hot Corrosion is another form of degradation which involves both oxidation and sulfidation in the presence of a molten salt on the substrate. This kind of degradation is very common in fossil-fueled power plants and aerospace gas turbines. In a coal-based power plant, burning of coal leads to the formation of ash. This ash may contain various combustion products present in the coal. Coal has usual impurities of sulfur, salts and silica particles. Ash formed after burning of coal travels towards cooler parts of the superheater tubes and deposits there. The deposition which occurs at very high speed results in partial erosion of the tubes. Ash deposit is usually solid at the time of deposition as the temperature of heat exchanger tubes is usually less than 540°C. However, with time the salt and sulfur, which have already reacted to form a sodium sulfate, react with more sulfur dioxide and sulfur trioxide, forming complex sulfates of sodium, leading to reduction in their melting point, at around $400°C$. This results in melting of the ash deposit and from this point, suddenly the corrosion rate starts increasing very fast. This increase in corrosion rate is due to the presence of hot melt between the substrate steel and gaseous environment, leading to faster diffusion of gases and therefore resulting in faster corrosion/oxidation of the base alloy, steel. The whole sequence of hot corrosion process described above is given as follows.

Sulfur impurity in the coal reacts with oxygen forming SO_2 and SO_3 as follows:

$$S(g) + O_2(g) = SO_2(g), \tag{13}$$

$$SO_2(g) + \frac{1}{2}O_2(g) = SO_3(g). \tag{14}$$

Both these reactions have the possibility to take place at the temperatures and oxygen gas availability in the system. At certain times of lean power requirements, oxygen level is reduced and there is possibility of changing the conditions from oxidative to reducive. Under these sub-stoichiometric conditions, formation of H_2S takes place, which is quite corrosive and can cause severe corrosion to the boiler tubes by direct reaction, forming stable compounds such as iron sulfides. This corrosion process is of several order magnitude higher than the corrosion in SO_2 environment.

Salts, mainly in the form of NaCl and KCl, react with SO_2/SO_3 to form Na_2SO_4 and K_2SO_4. Both these products are solid at the temperature of boiler operation and hence remain on the boiler tubes (superheated tubes) as solid particles. Under these conditions, there is no corrosion to the boiler tube. However, further reactions of these salts with SO_2 and SO_3 as per following reactions,

$$K_2SO_4(s) + SO_2/SO_3(g) = K_2S_2O_7(s) \quad (407°C, \ 150 \ \text{ppm} \ SO_3), \tag{15}$$

results in a eutectic mixture of $K_2SO_4 + K_2S_2O_7$ to form a mixture which brings down the temperature of combined products below

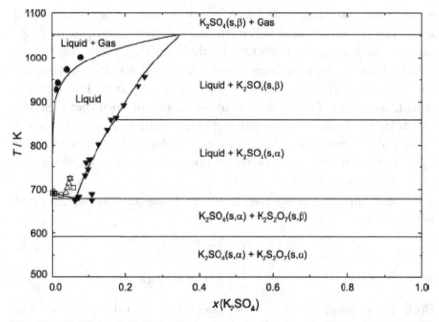

Fig. 12. K_2SO_4 and $K_2S_2O_7$ Phase diagram showing the Eutectic at $\sim$683 K (410°C).[12]

410°C and hence turns the deposited salt into liquid as shown in the phase diagram of K_2SO_4 and $K_2S_2O_7$ (Fig. 12).

By such a mechanism, the pyrosulfate can react aggressively with any protective iron oxide scale on the tubes, and lead to accelerated wastage through fluxing of the oxides and attack of the substrate metal.

$$K_2S_2O_7 + 3Fe = Fe_2O_3 + K_2SO_4. \tag{16}$$

The corresponding sodium system can become liquid at 400°C with about 2500 ppm of SO_3.[13] Such high concentrations of SO_3 is possible in the stagnant regions beneath deposits, so a similar attack by $Na_2S_2O_7$ may occur when a high-sulfur coal produces combustion gases containing high levels of sulfur oxides.

$$Na_2SO_4(s) + SO_2/SO_3(g) = Na_2S_2O_7(s) \quad (400°C, \ 2500 \ ppm \ SO_3). \tag{17}$$

Reid[14] has pointed, however, that the levels of SO_3 present at this location in a boiler, burning with a typical coal, are such that $K_2S_2O_7$ is unlikely to be found at temperatures above about 510°C, and $Na_2S_2O_7$ only up to about 400°C.

There is another form of hot corrosion that takes place in aerospace gas turbines, especially in those planes flying close to sea or stationed on warships or protecting sea boundaries. The air laden with salt enters the combustor, reacts with sulfur, a usual impurity present in the aviation fuel, forming sodium sulfate. Another impurity often present in the aviation fuel is molybdenum and/or vanadium, which in turn react with sodium sulfate to form low melting compounds, which melt immediately as the gas turbine temperature is always more than 900°C, which is above the melting temperature of sodium sulfate and sodium molybdate or vanadate. The details on this hot corrosion can be seen in Ref. 1.

8. Defect Structure and Diffusion

The process of oxidation occurs initially by the direct reaction of oxygen with the metal forming thin monolayer by chemisorption and subsequent layers forming a thin layer. This oxidation may stop here if the temperature is low or if the oxygen partial pressure is lower. Under these conditions, as discussed earlier, the kinetics follow logarithmic law. However, at atmospheric pressures and high temperatures, such thin layers are formed very quickly and for further oxidation, either the gaseous ions need to diffuse through the initially formed thin layer to reach the oxide/metal interface or metal ions need to diffuse from the metal/oxide interface to react with oxygen gas at the oxide gas interface. The diffusion process basically depends upon the defect structure of the particular oxide formed. Thus, it is important to know the defect structure of the initially formed thin layer oxide in terms of number of defects, such as point defects and their growth rate with temperature. Further, the grain size of the oxide is also important. Faster diffusion occurs through smaller grains than the larger grains.

The fact that oxide defect structure is really important in determining oxidation rate has been confirmed from the Wagner–Hauffe rules, according to which if the defect concentration is changed by addition of a small impurity of another element of different oxidation state, it would change the defect concentration of the oxide. In order to explain this effect, the oxides are divided into two categories, p-type and n-type, based upon the migration of electron holes or electron, respectively. An oxide such as ZnO is an n-type of oxide with excess metal defects such as Zn interstitials. If a very small impurity of a low oxidation state element such as Li^+ is added, it will enhance the number of interstitials defects and also the current carrying electrons and hence increase the oxidation rate of zinc. However, if an element such as Cr^{+++} is added, it would reduce the effective interstitials and hence number of current carriers, the electrons, and oxidation rate would decrease. A similar example is that of FeO, which is p-type oxide and has iron vacancies. When a Li^+ ion is added, the number of iron vacancies are reduced and oxidation rate is reduced. Addition of an element of higher oxidation state such as Cr^{+++} results in creating more vacancies and thus higher number of current carrying species electron holes and hence oxidation rate increases.

The effect of grain size is very important. It must be recognized that diffusion of charge carrying species such as ions can take place via lattice or bulk diffusion through the crystallite grain or it can follow the path through the boundary between two grains. The latter is relatively easy path due to weak van der Waals forces, joining the two grains, while lattice diffusion requires higher energy to move across the lattice through point defects. Thus, it is recognized that at lower temperatures, the grain boundary diffusion is preferred, while at higher temperatures, bulk diffusion is the main mechanism of diffusion. At many other temperatures, the diffusion usually occurs by both the mechanisms. Further, many high temperature alloys are designed by keeping smaller grain size as it helps in faster diffusion of selective oxide forming element, thereby forming the protective oxide faster, compared to an alloy of larger grain size.

9. Stresses in Oxide

Oxide growth process takes place by either diffusion of metal ions through the oxide or oxygen ions in the reverse direction. There are several factors which generate stresses during this diffusion process. Some of these factors are as follows:

(a) Pilling–Bedworth ratio — This is due to the difference in the molar volume of oxide to the metal. Whenever a new oxide is formed, its volume can be either less or more than the initial molar volume of the metal. The ratio of two is called Pilling–Bedworth ratio. In case this ratio is less than one, the oxide accumulates compressive stresses, while the P.B. ratio is more than one, stresses are tensile in nature.

(b) Epitaxial Stresses — These are mainly due to misalignment or mismatch between scale and metal. Small change in dimensions of the grains or a particular plane can cause severe stresses.

(c) Geometry of Surfaces — This is a factor which causes stresses because the diffusion process can be modified, based on whether the surface is flat, concave or convex. Thus, the geometry of a surface with the type of diffusing ion and its molar volume will generate stresses which can be tensile or compressive.

(d) Thermal Stresses — These are very important as frequent heating and cooling happens in many industrial components. Thus, when a metal component is cooled from a rather high temperature, it can generate stresses which are generally tensile in nature.

So what is the problem with these stresses? When a metal or alloy is oxidized, these stresses continually grow with time and keep accumulating. Under these conditions, the oxide may show protective nature. However, after a sufficient long time, the stress accumulation becomes so high that the oxide is unable to hold it and would like to release these accumulated stresses. And when the stresses are released, the protective oxide can form a micro-crack or over a longer period, can lead to oxide spalling, thus leading the linear kinetics from the rather protective parabolic kinetics. This is basically an indication of the severe damage of the component from corrosion and it needs either repair or replacement of the component.

10. Active Element Effect

Spalling of the oxide layers is a serious problem for many high temperature alloys. In most of the cases, spalling occurs when the material at high temperature is cooled suddenly to low temperatures. This is mainly due to release of thermal stresses as explained in the above section. Thus to make the alloy sustain these thermal stresses, additional mechanisms are created in the alloy which hold the oxide layer even during thermal cycling. This is done by addition of a very small amount (0.1%) of rare earth elements such as yttrium, cerium or lanthanum. These elements are also known as active elements as they have stronger affinity towards oxygen. It is now well proven that the superalloys with 0.1% of Y, Ce, or La show scales which do not spall under thermal cycling conditions.

There are various theories on the active element effect as suggested by several workers:

(1) Promotion of nucleation of alumina, thereby reducing the aluminum content necessary to form continuous protective alumina layer in alumina forming superalloy.
(2) Formation of intermediate oxide layers, which act as barrier to outward diffusion of metal ions.
(3) Active elements act as sinks for voids, thereby enhancing scale adherence by avoiding void precipitation at the oxide/metal interface.
(4) Formation of oxide pegs which enhance the oxide adherence, thereby reducing spalling of oxide.
(5) Enhancement in the plasticity of the scale, so that the scale can deform without breaking during stress release process.
(6) Well known sulfur effect, which means presence of active elements can bind last traces of tramp sulfur, present in the oxide metal interface, thereby strengthening the bond between oxide and the metal.

Each theory has been found to be applicable on certain alloys under a specific condition, but may not be applicable on other alloys under similar conditions. That is why there is no universal theory of active element effect. However, theory such as pegging effect is well accepted for pinning the oxide scale to the substrate and barrier effect for lower oxidation rate. However, the more universal theory to explain the spalling resistance of the oxide scale comes from the latest theory, which is "Sulfur Effect". According to this theory, the bond weakening between the substrate and the oxide scale is due to the segregation of sulfur at the oxide/metal interface. Once this ppm level of sulfur is removed, scale adherence gets improved. Presence of a 0.1% of active element in the alloy, basically helps to remove this sulfur and hence improves the oxide adherence. This has been well proven from various experimental observations of testing oxide adherence from sulfur free alloys and those with ppm level of sulfur impurity.[1]

11. Selection Criteria for the Oxidation Resistant Materials for High Temperature Application

Thousands of components in different industries are exposed to high temperatures as well as severe corrosive environments. Chemical process industry, refineries, petrochemical plants, power plants are a few examples to cite. Safety of plant and the personnel working there is the most serious concern of the authorities. Hence, material selection is of utmost importance in order to save the plant from unwanted accidents. The most important selection criteria must take care of retention of adequate strength at the temperature of operation and ensure there is no sudden reaction of the material with environment.

The basic requirements of materials, operating at high temperatures, can be listed as follows:

(1) High Melting Point.
(2) Microstructure Stability.
(3) Sustained mechanical properties such as UTS, creep and fatigue strength.
(4) Low corrosion rate with the environment to which it is exposed.

The high melting temperature is required as it helps to use the materials at higher temperature. It is also important from the requirement of creep resistance. It is well known that creep in material starts above 2/3 of its M.P. Thus, the higher the M.P., the higher would be the temperature it can be used. High temperature materials are required to have same microstructure as change in microstructure changes the diffusion characteristics, and hence the oxidation behavior. The high temperature material must have strong strengthening elements which restrict the movement of dislocation.

Once the basic physical and mechanical properties requirement is taken care, material selection becomes straightforward for high temperature oxidation resistance. From oxidation point of view, a slow growing chromia or alumina oxide scale should form to protect the alloy from further oxidation. For this, the material composition should be such that a selective oxide scale should form within a very short time of exposure to the high temperature environment. Further, the concentration of protective oxide forming elements, such as Cr or Al should be such that it does not deplete considerably after the oxide formation, thereby leading to inferior or mixed oxide formation. Thus, for stainless steels and superalloy oxidation, following criteria are sufficient:

(1) For stainless steels, the Cr concentration should be between 20% and 22% for a sustained chromia oxide scale formation for a sufficient long duration.

(2) For Ni-based superalloys, the Cr concentration should be above 10%. Superalloys between 10% and 15% Cr can show a sustained chromia scale formation.

(3) For alumina forming superalloys, minimum Cr and Al concentration required is 10% and 5%, respectively. For higher Cr concentration, slightly lower Al concentration can be acceptable.

(4) The temperature limit for chromia forming alloys from oxidation point of view is 900°C as above this temperature chromia scale gets de-stabilized due to formation of volatile Cr(VI) oxide. For temperatures above 900°C, it is recommended to use alumina forming alloys.

(5) For sulfidation resistance, the alloying composition must be rich in Cr. For example, a Ni-base superalloy with 10% Cr is very good for oxidation resistance, but bad for sulfidation resistance. In order to make it sulfidation resistance, Cr level should be above 12–15%. For alumina forming alloys also, higher Cr in the alloy helps in sustained sulfidation resistance.

Let us take some typical examples of material selection in some industries:

(a) Fossil fuel power plants

There are two important criteria for material selection for a coal-based power plant. Heat exchanger tubes selected will be suitable for boiling of water to form steam and the superheated steam can then run the turbine. For former, usually a low alloy steel such as 21/4Cr–1Mo is excellent as it is able to protect the inner part of the tube from the water corrosion, provided the water chemistry is suitably controlled. The inner wall temperature of the boiler tube ranges from 250°C to 300°C. The low alloy steel provides good oxidation resistance by forming a good magnetite layer. The temperature on the superheated tubes towards steam side is rather high, in the range of 450–540°C for which 21/4Cr–1Mo is not a suitable material, thus a steel of type 9Cr–1Mo is used, which provides excellent oxidation resistance to the inner side of the superheated steam tube.

However, the outer part of the tube is exposed to a very complex environment consisting of oxidation, sulfidation and hot corrosion. This environment depends primarily on the impurities in coal such as sulfur, salts or silica. With ppm level of sulfur and silica, it is possible to have an environment which can cause sulfidation and also deposition of ash leading to hot corrosion. The only way to avoid this is by the application of a chromium rich coating which can resist the sulfidation and hot corrosion. Thus, methods like thermal spray coating of Ni–25Cr or laser cladding with Ni–25Cr powder, can prevent corrosion from fire side in a boiler.

(b) Material Selection for Gas Turbines

Gas turbine is another example where material selection is based upon the high temperature strength as well as on the oxidation resistance of the alloy. The most critical component in a gas turbine is the turbine blade which is exposed to a temperature above 1400°C in a highly oxidizing environment, and possibility of sulfidation and hot corrosion in case the aeroplane lands and takes off from an offshore field. The high temperature requires rather highly stabilized material which can sustain strength at a temperature above 1200°C. For this, mostly alumina forming superalloys highly strengthened by host of substitutional elements such as Mo, W, Ti, Zr, B, and V and also by intermetallic precipitates such as γ' and γ''. However, the biggest problem of such an alloy is that it has lower melting temperature (less than 1200°C) and also has the problem of segregation of heavy elements during heat treatment. If however, a single crystalline material of these alloys is used, in such case some alloying elements such as V, Zr, B are removed, resulting in an increased melting temperature, greater than 1300°C. Even after that, unless the alloy is properly coated by a thermal barrier coating and further cooled by some holes in blades, it is not possible to use it for turbine blade application.

12. Conclusion

High temperature corrosion and oxidation is a serious problem for many industrial components operating at high temperatures. Thus, one of the most important requirements is the proper material selection strategy. This basically depends on three important factors: sustainability of strength at the temperature of operation, microstructural stability and finally its reactivity to the environment. As far as structure stability is concerned, many high temperature materials used at high temperatures have close packed austenitic structure (FCC) and sustained high temperature strength is obtained from a combination of solid solution strengthening elements and precipitation hardening approach. For corrosion resistance, use of chromium and aluminum is made as protective oxide forming elements which result in the formation of thin, slow growing and adherent oxide. This also requires use of active element effect and also alloy development in a way to have minimum stress accumulation.

References

1. A.S. Khanna (Ed.), *Introduction to High Temperature Oxidation and Corrosion* (ASM International, Ohio, 2002).
2. P.B. Anderson and O.J. Krudtaa, Oxidation of titanium in the temperature range 800–1200°C, *J. Less Common Metals* **3**(2) (1961), pp. 89–97.
3. B. Cox, Oxidation and corrosion of zirconium and its alloys, *Corrosion* **16**(8) (1960), pp. 380–384.
4. A.C. Rion, D.F. Cowgill and B.H. Nilson, Review of the oxidation rate of zirconium alloys, *Sandia National Laboratories*, SAND2005 – 6006, 2005.
5. Stainless steel information centre. Retrieved from http://www.ssina.com/composition/temperature.html.
6. *Corrosion Handbook for Stainless Steel*, 10th edn., Outokumpo Oyj, (pub.) Finland, 2009.
7. Mathew J. Donachie and Stephen J. Donachie, Superalloys a Technical Guide, ASM International Ohio, USA, 2002.
8. Stephen Lamb, (Technical editor), CASTI Handbook of Stainless Steels and Nickel Alloys, ASM International, Ohio, USA, 2004.
9. A.S. Khanna, High Temperature degradation of materials, in "Handbook of Environmental Degradation of Materials", Myer Kutz (Edt.), William Andrews Pub. USA, 2005.
10. R.E. Smallman and R. J. Bishop, in *Modern Physical Metallurgy and Material Engineering*, J.B. Ray (Ed.), Elesvier, (1999), p. 64.
11. G.Y. Lai, *High Temperature Corrosion of Engineering Alloys* (ASM International, Ohio, 1990), pp. 117–143.
12. M.K.W. Wang, Study of the thermochemistry for oxygen production for a solar sulfur-ammonia. Electronic Thesis (2012). Retrieved from https://escholarship.org/uc/item/5464k3bn.
13. A.S. Khanna, Fireside corrosion and erosion problems in coal based power plants, *National Workshop on Boiler Corrosion*, 11, 12th April, (1995), NhML Jamshedpur, India.
14. W.T. Reid, *External Corrosion and Deposits — Boilers and Gas Turbines* (Elsevier, New York, 1971).

Chapter 2

Degradation of Mechanical Properties of Materials at High Temperatures in Corrosive Environments

A.S. Khanna

Department of Metallurgical Engineering and Materials Science
Indian Institute of Technology
Mumbai 400076, India

1. Introduction

One of the important criteria for material selection is its mechanical properties, viz., tensile and yield strength, fatigue strength and if it is operating at high temperatures, degradation due to creep also becomes very important. Going into details, it is found that while selecting materials for high temperature applications, it is not the room temperature strength, but the sustaining strength at the required temperature that is important. This means that if a material is selected for operation at 500°C, its strength at this temperature must be considered rather than its strength at room temperature. This is because materials lose their strength at high temperatures due to high atomic mobility, which creates more defects and hence easy diffusion paths for dislocation movement, thereby losing strength. Therefore, in order to select materials to withstand higher temperatures, it is necessary to choose the minimum strength required at the temperature of operation. Figure 1 gives the tensile strength of various steels as a function of temperature, indicating that all the steels lose strength with increase in temperature. It is, however, interesting to note that the extent of strength loss is different for various steels. A simple low carbon steel loses its

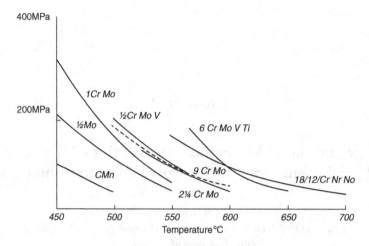

Fig. 1.   Effect of temperature on the tensile strength of various steels.

strength almost to a very low value of about 20 MPa at 500°C, while
steels with elements such as Mo, and/or V, or Ti, the strength loss
is substantially reduced and such steels sustain better strength at
higher temperatures.

What could be the reason for various steels losing strength at a
different pace as the temperature increases? This is basically due to
the formation of several obstacles, which resist the dislocation movement and hence sustain higher strength with increasing temperature.
Formation of various obstacles depends on the steel composition.
Presence of interstitial and substitutional solid solutioning elements
create elastic stress waves which resist the dislocation movement.[1] In
the same way, creation of precipitates within the grain structure does
impede dislocation movement. Presence of elements such as Mo, Nb
in steel acts as substitutional solid solution strengtheners (SSS) and
hence helps in retaining strength of steel for higher temperatures. In
the same way, presence of precipitates, such as carbides in steel such
as iron carbide, molybdenum carbide, further helps in strengthening
various steels.

Figure 2 shows the variation of a very important mechanical
property, creep rupture strength as a function of temperature. Like
tensile strength, this too decreases considerably with increase in

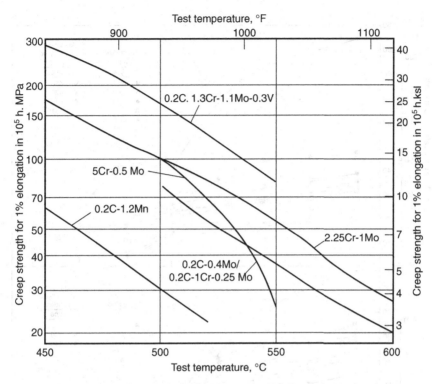

Fig. 2. Comparison of various steels in terms of their variation in their creep rupture life as a function of temperature.[2]

temperature and the trend for simple carbon steel versus modified steels with Mo, Nb, or V, varies in a similar manner.

A similar kind of situation can be seen in other high temperature alloys such as austenitic stainless steels and superalloys. A comparative variation of tensile strength and yield strength with temperature for various 200 and 300 series stainless steels is given in Fig. 3. It can be seen that several high temperature alloys retain strength when strong SSS elements such as Mo, Ti, Ta, Nb are added into these alloys. Further strengthening can be achieved by creating very strong precipitates such as intermetallic precipitates, for example, Ni_3Al (Υ'), $NiNb_5$ (Υ''), or Ni_3Ti (η) or inert oxide precipitates such as yttria, ceria. The former help in sustaining strength of superalloys up to 1050°C, while the latter help to sustain up to 1150°C. Figure 3

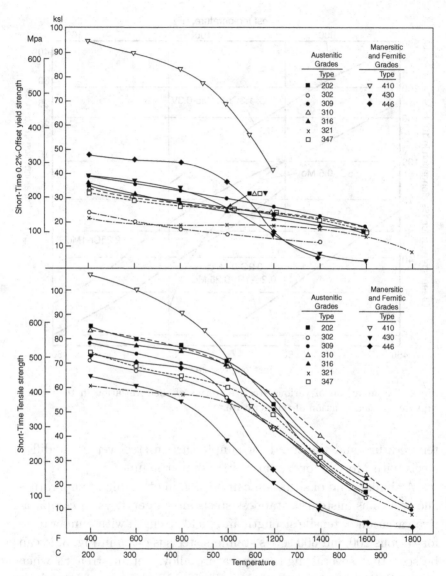

Fig. 3. Short term tensile strength as well as 0.2% yield strength for various
stainless steels versus temperature.[3]

gives the variation of tensile strength of various superalloys as a
function of temperature. It can be seen that many alloys such
as Inconel 718, X-750, and Nimonic 80A, 115, and 90, which are
strengthened by various intermetallic precipitates, retain strength

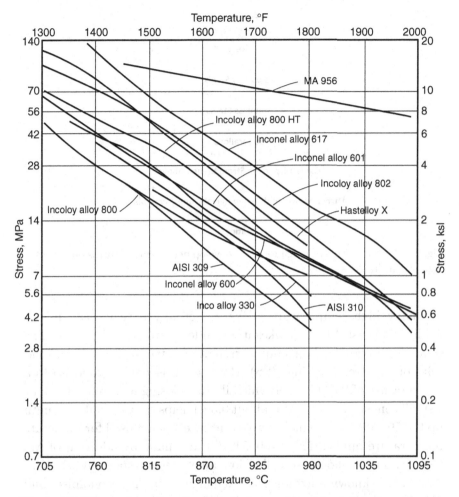

Fig. 4. 10,000h rupture strength of selected wrought solid solution strengthened nickel base superalloys versus temperatures. Oxide strengthened MA 6000 has been included as comparison.[4]

till 1000°C, while Nimonic MA 6000, which is strengthened by inert oxides such as yttria, sustain strength till 1150°C.

Figure 5 gives a systematic scheme of development of high temperature alloys by simply changing the alloy composition to enhance strengthening or corrosion resistance. It can be seen that the alloy microstructure and the fabrication route also helps in retaining best mechanical properties of the alloys. Iron, which is the base metal

Fig. 5. A systematic approach for alloy development to make corrosion resistant high strength alloys.

for many high temperature alloys, is actually a very weak metal. Addition of just 0.1% C makes it a wonder material called steel. The small addition of carbon simply strengthens steel by interstitial solid solutioning strengthening (ISS). However, this steel is good up to a temperature of 250°C, after which it starts losing its strength. Thus, a small concentration of Mo (about 0.5%) helps to sustain its strength up to 350–400°C. The most important steels being used for moderate temperature applications (400–540°C) are made by addition of Mo and Cr. These alloys are called low alloy steels, especially 2¼Cr–1Mo, a well-known alloy for heat exchanger tubes for economiser and 9Cr–1Mo steel for superheater tubes. Presence of chromium helps in forming an iron chromium spinel oxide instead of pure iron oxides, while Mo helps in SSS and formation of Mo_2C and therefore, helps in strengthening steel and sustaining its strength up to 540°C. A small amount of Nb and V also helps in sustaining this strength, the latter also helps in grain refinement.

This kind of alloy modification can only enhance the temperature capability of alloy up to 600°C maximum, that too by addition of other alloying elements such as W and Nb. Such modified 9Cr–Mo–W alloys are used for supercritical power plants for a pressure of about 300 bar and temperature of 620°C. Further enhancement of high temperature capability can be achieved only by changing the microstructure. Thus, the next step is to change the ferritic structure of the steel to austenitic, which enhances the capability of austenitic stainless steels to withstand 750°C. Austenitic structure, being close packed FCC, is able to sustain suitable strength at this temperature, but needs help of additional SSS elements such as Mo, Ti, Nb and Ta for sustaining strength at still higher temperatures. Such stainless steels are called high alloy stainless steels. Alloy 904 is one such stainless steel which can sustain strength till 900°C.

For further strength sustainment, the SSS or ISS effect is not enough. One needs strong strengthening precipitates such as intermetallic precipitates or inert precipitate dispersion to achieve suitable strength till 1050 and 1150°C, respectively. These alloys are respectively called as superalloys and oxide dispersion strengthened alloys.

Beyond 1150°C, it is not possible to enhance the high temperature capability of the materials just by alloying alone. Further modification can be achieved by changing the method of fabrication, for example, powder metallurgy or investment casting. Thus by modifying the method of casting from a conventional casting to directionally solidified casting or single crystalline casting and thereby eliminating some alloying elements, required for grain boundary strengthening and grain refining, it is possible to enhance the alloy capability to 1220–1250°C.[5]

Thus, it is clear that the alloy strengthening and sustaining the strength at various temperatures require making use of all possible strengthening mechanisms and also usefulness of specific alloying elements. The schematic of various alloy development with this strategy is depicted in Fig. 5.

It is now important to consider how the degradation in mechanical properties can be further affected when the material is being exposed not only to high temperature but also to a corrosive environment such as air, oxygen, SO_2, H_2S, etc. Let us now deal with detailed treatment of two most important properties at high temperatures, viz., creep and fatigue.

2. Creep

Creep is, perhaps, the most important property of materials, operating at high temperatures. It is defined as the slow deformation of material under constant stress at a particular temperature. Figure 6 shows a standard creep curve, indicating three stages of deformation. Stage one is called primary creep, where the strain rate increases with time, mainly due to the work hardening. Secondary creep is the next one, where there is an equilibrium between the work hardening and stress relieving process. This happens due to the dislocation glide mechanism, due to which impeded dislocations move, easing strain hardening. Secondary creep is the temperature dependent process: higher the temperature, higher is the creep rate and in the same way at higher applied stress, the creep rate is higher. Secondary creep is the most important stage of creep. This value is used to design high temperature components. The secondary creep is also called as steady state creep or diffusional creep. The minimum secondary creep rate is of most interest to design engineers since failure avoidance is commonly required and in this area, some predictability is possible.

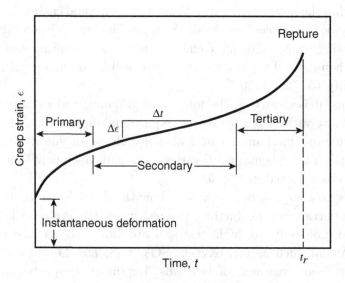

Fig. 6.   A typical creep curve.

In general, two Standards are commonly used[1]:

- The stress to produce a creep rate of 0.0001% per hour (1% in 10,000 hours).
- The stress to produce a creep rate of 0.00001% per hour (1% in 100,000 hours or approximately 11.5 years).

The first requirement would be typical of that for gas turbine blades, while the second standard would be most typically seen in applications such as those for steam turbines.

With elapsing of time, the material, during secondary creep, starts getting deteriorated due to formation of voids and at a certain time, when precipitated at grain boundaries, can cause grain boundary sliding. This is the time when secondary creep tends to transform to tertiary creep, beyond which the material can fail at any instant. Figure 7 shows how with time the material develops voids, whose concentration increases with time and finally leads to grain boundary sliding.

One of the most popularly used techniques in representing creep rupture data is Larson–Miller time–temperature parameters. This parameter can be derived from the stress and temperature dependence of the creep rate or time to rupture as $P_{LM} = T\,[21.577 + \log(t_r)]$. Figure 8 gives a typical representation of life prediction using this approach parameter. This data can be used for rupture life prediction for many industrial components.

Since for design, the secondary creep is the most important parameter, it would be discussed in detail here. Secondary creep is a thermally activated process, the minimum secondary creep rate can be described by a fundamental Arrhenius equation of the form,

$$\dot{\varepsilon} = \frac{d\varepsilon}{dt} = K\sigma^n \exp\left(-\frac{Q}{RT}\right),$$

where T is temperature [K], K is creep constant, n is stress exponent (varying from metal to metal, most common value is $n = 5$; determined experimentally by plotting the strain rate as a function

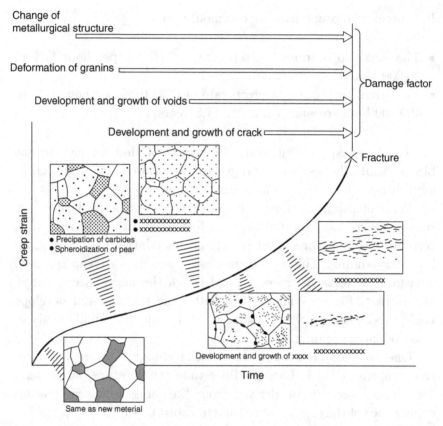

Fig. 7.   Change of microstructure with time in the secondary creep regime.[7]

of stress), σ is applied stress/nominal stress, Q is activation energy for creep, and R is gas constant (8,314 J/mol K).

2.1 *Corrosion and Creep*

The above equation for steady state creep is valid when the creep test is carried out in vacuum or in protected environment such as argon gas or helium. What would happen to material when exposed to a corrosive environment such as air, sulfur bearing gases or carbon dioxide? One of the additional effects the metal would face would be oxidation and corrosion of the surface, which can cause several

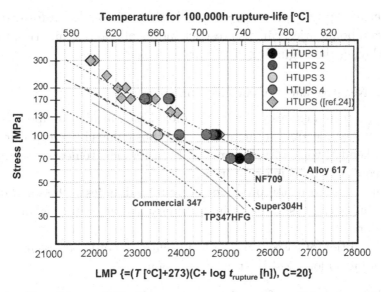

Fig. 8. Plot of Larson–Miller Parameter versus stress for a number of commercial alloys.[7]

changes in the material which can either increase or decrease the creep rate. Let us look into both these cases.

2.1.1 Positive effect of corrosion on creep

- When a metal undergoing creep is exposed to oxidizing environment such as oxygen, it can form an oxide which can protect the metal from further corrosion and thus help in enhancing creep resistance. If a very stable oxide is formed, it controls oxide growth and hence restricts weight loss and thus results in containment of member diameter.
- Such a surface oxide formation can also restrict the multiplication of dislocation during slip (Frank–Read source), as the oxide is formed on surface, fixes one edge, which restricts the dislocation multiplication and hence restricts the slipping process and thus helps in enhancing creep resistance.
- Formation of internal oxides due to oxygen dissolution, if it takes place within grains, results in higher strength due to precipitation hardening and therefore results in better creep resistance.

- Certain specific products such as borides, zirconides, if formed at grain boundaries, can also restrict sliding of grain boundaries, thus delaying the beginning of tertiary creep.

2.1.2 Negative effect of corrosion on creep

- Unprotected oxides formed due to oxidation at high temperatures may spall, leading to metal loss and thus enhancing creep rate.
- Kirkendall voids formed can lead to faster achievement of tertiary creep, thus reducing rupture life.
- Unwanted corrosion products at grain boundaries such as sulfides, carbides can result in grain boundary sliding, resulting in early failure.
- Dissolution of gases can lead to formation of ordered precipitates, leading to localized brittleness and failure.

In case, the oxide formed due to oxidation is fast growing, leading to spalling, this would help in reducing the thickness of the metal, thus decreasing its load bearing capability. The continuous decrease in metal thickness due to spalling will reduce the area of cross-section, resulting in decrease in creep resistance. This in turn will modify the steady state creep as shown below.

Let us take a specific example where loss in thickness due to oxidation of a member undergoing creep affects the creep rate. Let us take cylindrical member of thickness "do" which loses a thickness of "x" after oxidation.

The steady state creep rate of such a system can be written as

$$\dot{\varepsilon} = A\sigma^n \exp\left(-\frac{\Delta E}{RT}\right).$$

At constant temperature, the terms within the parenthesis is constant and can be indicated with a new constant A' in combination with earlier constant "A". The equation now looks as

$$\dot{\varepsilon} = A'\sigma^n.$$

In this equation, creep rate is proportional to the applied stress σ, which in turn is nothing but F/A, where F is applied load and A is area of cross-section of the member. Area of a round member, such as a rod, can be written as πr^2, where r is radius of the member. This can also be written as $\pi(d_o/2)^2$, where d_o is the diameter of the rod. Now, the stress σ is proportional to F/A or $F/\pi(d_o/2)^2$. Or it can be written as

$$\dot{\varepsilon} = A\sigma^n = \frac{A1}{d_o^2},$$

where d_o is the initial diameter of the rod. Since this rod is exposed to corrosive environment, say air, it may lead to the oxidation of rod, oxidation may lead to formation of oxide, whose growth rate can be expressed as

$$x = Kt^m,$$

where x is the thickness of oxide formed, in time t, m is the exponent which describes the kinetics of oxidation. In case $m = 1$, the kinetics are linear, $m = 2$ means parabolic and $m = 3$, cubic kinetics and so on. K is the corresponding rate constant. As the oxidation occurs, loss in thickness due to oxide can result is the effective thickness of the rod. This means if the thickness of the oxide is x in time t hours, the effective thickness remaining of the rod can be $d_o - 2x$

Thus, the change in stress due to oxidation can be written as

$$\Delta\sigma\alpha\left\{\frac{1}{d_o^2} - \frac{1}{(d_o - 2x)^2}\right\}^n,$$

$$\alpha\left\{\frac{1}{d_o} - \frac{1}{(d_o - 2x)}\right\}^{2n},$$

$$\alpha\left\{1 - \frac{1}{(d_o - 2x)}\Big/d_o\right\}^{2n}.$$

Using the binomial expansion and neglecting higher terms, and replacing x with Kt^m, where K is the rate constant of oxidation and m is the exponent of time, which indicates the kinetics of oxidation, linear for 1 and parabolic for 2 and so on, we obtain

$$\Delta\sigma\alpha\left\{\frac{1 - 2Kt^m}{do}\right\}^{2n}.$$

Now, putting the value of stress in steady state creep equation, we get the steady state creep as

$$\dot{\varepsilon} = A \left\{ \frac{1 - 2Kt^m}{do} \right\}^{2n}.$$

Now, we can write a more general creep equation defining a term $\varepsilon_{(s,t)}$ as steady state creep in an environment (s) at a time (t) and $\varepsilon_{(t)}$ the steady state creep at time t without any environment, as follows:

$$\varepsilon(s, t) = \varepsilon(t) + \varepsilon(s, 0) \int_0^t \left\{ \frac{1 - 2Kt^m}{do} \right\}^{2n} dt.$$

This means when a material is under creep and also exposed to a corrosive environment, its creep rate differs by a factor given by the last term in the above equation, which in turn depends upon the rate constant and type of corrosion kinetics.

The effect of oxidation on creep for stainless steel is shown in Fig. 9. In Fig. 9(a), it is shown that preoxidation of steel leads to significant loss in rupture strength as oxidation can lead to scale formation, Kirkendal voids, etc., as described above. In Fig. 9(b), it is clearly evident that as the Ar gas environment is reduced from 1 bar to 10^{-3} bar, and the rupture strength falls rapidly with decrease in Ar pressure. Lower Ar pressure means more oxygen and hence

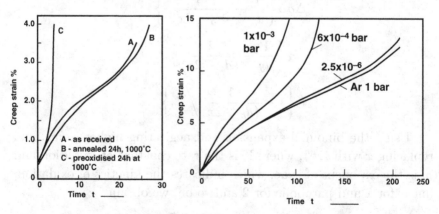

Fig. 9. Effect of oxidation on the creep strain of stainless steel oxidized at 1000°C: (a) effect of pre-oxidation and (b) effect of reducing pressure of protective gas Ar.

higher oxidation, leading to higher creep rate and hence lower rupture strength.

3. Fatigue

The deterioration of material under cyclic load is one of the serious problems. The material, when exposed to cyclic loads, results in the formation of intrusions and protrusions which at a later stage become the initiation points for the stress concentration and hence point source for crack initiation. Fatigue is usually measured in terms of fatigue strength, obtained from an SN curve, that is the stress versus number of cycles curve as shown in Fig. 10 for ferrous materials such as steel, the constant stress, below which the material cannot fail even after infinite number of cycles, is known as endurance limit. In case of non-ferrous materials such aluminum, where the SN curve keeps varying with number of cycles, the fatigue strength is the stress at a fixed number of cycles usually between 10^6 and 10^8 cycles.

The fatigue failure occurs usually by cleavage. The fracture surface appears bright as shown in Fig. 11(a) and if seen at higher magnification, one can count the number of cycles after which the material fails from the striations formed, as shown in Fig. 11(b).

The fatigue strength of material depends upon the type of loading which in turn decides the number of cycles at which the material

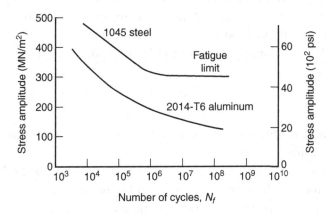

Fig. 10.   Stress versus number of cycle curve for steel and aluminum.

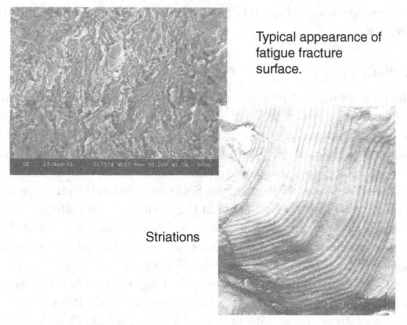

Typical appearance of
fatigue fracture
surface.

Striations

Fig. 11.   (a) A typical fatigue failure microstructure showing cleavage fracture
and (b) at higher magnification, showing striations.

fails. Based upon this, it is classified as high cycle fatigue (HCF) or
low cycle fatigue (LCF). HCF is defined as that in which the stresses
that cause deformation in each cycle remains within the elastic range.
Usually, failures under such conditions are expected only when the
stresses cross the yield stress.

The HCF is usually represented by Basquin equation:

$$N\sigma_a^p = C,$$

where p and C are empirical constant, σ_a is the stress amplitude and
N is the number of cycles.

LCF is actually more severe. The main definition of LCF is not
that the material fails here at lesser number of cycles, 10^4 or so, but it
is because of the fact that the stress change in each cycle is so severe
that each cycle is represented by a definite plastic deformation. This

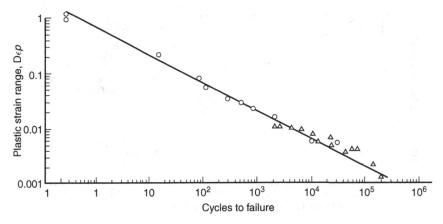

Fig. 12.   Coffin–Manson Plot for the LCF.[1]

is shown by Coffin–Manson relationship as given below:

$$\frac{\Delta \gamma_p}{2} = \gamma_f'(2N)^c,$$

where $\frac{\Delta \gamma_p}{2}$ is the plastic strain amplitude, γ_f' is the fatigue ductility coefficient defined by the strain intercept at 2N = 1, 2N is the number of strain reversed to failure (1 cycle = 2 reversals), c is the fatigue ductility exponent which varies between 0.5 and 0.7, a smaller value of C results in larger value of fatigue life.

The plot between $\Delta \gamma_p$ versus $2N$ is known as Coffin–Manson plot (Fig. 12).

3.1 *HTF in Corrosive Environments*

Fatigue under corrosive environment is called corrosion fatigue, which refers to any cyclic loading in a corrosive environment, such as cyclic loading in sea water, or other corrosive environments. However, high temperature fatigue has a different aspect. The Coffin–Manson plot as shown in Fig. 12 above is showing LCF as a function of number of cycles in a protected environment such as vacuum or inert environment. However, this variation changes to a double slope plot when the environment becomes corrosive such as air or a corrosive gas such as oxygen or sulfur dioxide as shown in Fig. 13. This

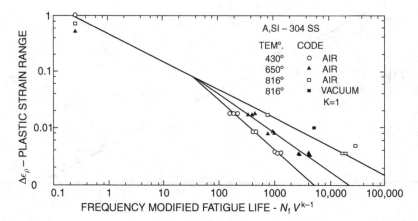

Fig. 13.   Double slope Coffin–Manson plot for the LCF of AIST Stainless steel
type 304 in vacuum and at three temperatures, 430°C, 650°C, and 816°C.

shows the Coffin–Manson plot for 304 stainless steel at various
temperatures, 430°C, 650°C, and 816°C. It can be seen that when the
test was carried out in vacuum, irrespective of temperature variation
from room temperature to 430°C, 650°C, and 816°C, all the points
fell on a single line, represented by a single slope Coffin–Manson plot,
but when the environment changed to oxygen, it is showing double
slope plot with a variation in slope at each temperature, indicating
higher effect at higher temperature.

This mainly happens when the environment changes from vacuum to active such as air. During cyclic oxidation, the crack
tip, if formed, gets oxidized and changes its path, tilting from
original variation in plastic deformation versus number of cycle plot.
Figure 14 also shows the effect of temperature on the LCF of stainless
steels. While in vacuum, the variation in plastic strain versus number
of cycles followed same plot, but in air, there was variation in plots
at each temperature, the higher temperature showing lower creep
resistance.

Another important effect of LCF is the frequency effect, as shown
in Fig. 15. It shows double slope Coffin–Manson plot when creep test
is carried out in a reactive gas such as oxygen while the points from

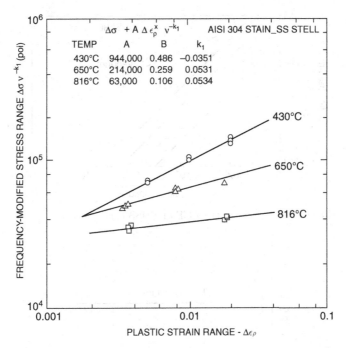

Fig. 14. LCF of AISI type 304 SS in air at three temperatures, 430°C, 650°C, and 816°C.

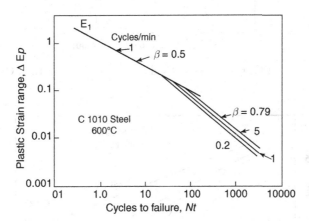

Fig. 15. LCF test of steel at various frequencies in vacuum as well as in air at various frequencies.

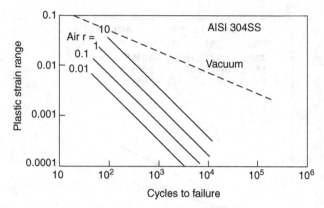

Fig. 16.   Effect of frequency on the LCF of 304SS carried out on vacuum and air.

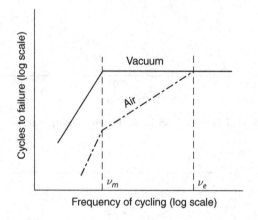

Fig. 17.   Effect of frequency and air environment on the LCF.

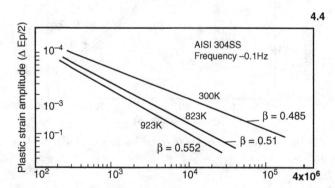

Fig. 18.   LCF of 304 SS at three different temperatures 300 K, 823 K, and 923 K at various frequencies, 0.485, 0.51, and 0.552, respectively.

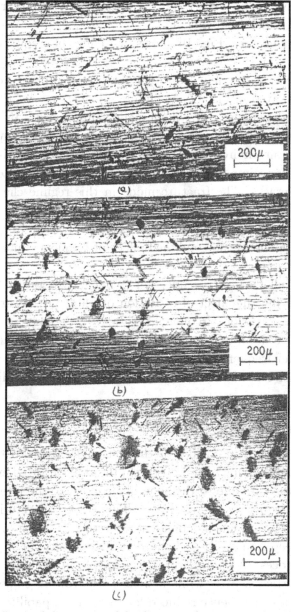

Fig. 19. Effects of frequency of loading on the extent of oxide formation frequencies are 5.6, 1, and 0.22 cpm and the test is at 593°C. At lower frequency, there is more chance for oxidation as the member is exposed to environment for longer time.

all frequencies fall at one straight line when the test is carried out in vacuum at various frequencies.

In the same way, Fig. 16 shows the effect of frequency on the LCF of 304 Stainless steel.

The effect of frequency on LCF at high temperatures can be explained as given in Fig. 17. At high frequencies $v > v_e$, the damage process is independent of frequency and failure occurs by the usual transgranular fatigue mechanism. In the intermediate range ($v_m < v < v_e$), an air environment is capable of interacting with fatigue cracks. In this regime, the environment is capable of interacting with the fatigue cracks. The fracture mode in this regime changes from transgranular to intercrystalline type. At lower frequencies, $v < v_e$, microstructural instabilities and intergranular fracture may result.

Figure 18 shows the combined effect of frequency and temperature for 304 stainless steel.

A pictorial view of the effect of frequency on the LCF can be shown by seeing the effect of oxidation on the crack decoration due to oxidation at various frequencies (Fig. 19). At lower frequencies, the cracks are decorated more and are also bigger compared to the same at lower frequency. This is because at lower frequency there is more time for crack to react with oxygen and hence it oxidizes more.

4. Summary

Oxidation of metals and loss of their strength leads to the degradation of metal at high temperatures. In addition, further material degradation occurs due to the synergistic effect of the interaction of mechanical properties and oxidation. Thus, creep of material at high temperature is very much dependent upon the oxidation of the metal or alloy. Creep rate can be enhanced or reduced depending upon the oxidation process. A strong protective oxide formation may affect the creep properties, a little while a weak, fast growing oxide will lead to material loss and hence reduced load bearing capability measured in terms of loss in area of cross-section, leading to higher creep rate. High temperature fatigue also is oxidation dependent. In vacuum, the fatigue strength is not affected at any temperature nor the frequency cycle has any effect. However, in a corrosive environment, fatigue strength deteriorates sharply by increasing temperature or by reducing cyclic frequency. Hence, while measuring creep of materials at higher temperature, additional effect of environment must not be neglected.

References

1. G. Dieter, *Mechanical Metallurgy* (McGraw Hill Book Company, 1988).
2. *ASM Handbook*, Vol. 1, Mechanical properties of steels at elevated temperatures, ASM International (pub.) Ohio, USA.
3. *High-Temperature Characteristics of Stainless Steels*, A Designers' Handbook Series No 9004 (NIDI).
4. F.L. VerSnyder, Superalloy technology — today and tomorrow, *Proc. High Tempertaure Alloys and Gas Turbines*, R. Brunetaud, D. Coutsouradis, T.B. Gibbons, Y. Lindblom, D.B. Meadowcroft and R. Stickler (eds.) (Belgium, 1982).
5. M. J. Donachie and S. J. Donchie, Chapter 12, *Superalloys: A Technical Guide* (ASM Publication, Ohio, 2002).
6. M.F. Ashby, A first report on deformation-mechanism maps, *Acta Metal. Mater.* **20** (1972), pp. 887–897.
7. E. A. Payzant, M. P. Brady, B. Yang and H. Wang, Creep-resistant, Al_2O_3-forming austenitic stainless steels, *Science* **316** (2007), pp. 433–436.

Chapter 3

Materials Development Aiming at High Temperature Strengthening — Steels, Superalloys to ODS Alloys

Shigeharu Ukai

Laboratory of Advanced High Temperature Materials
Research Group of Energy Materials
Division of Materials Science and Engineering
Hokkaido University, Japan
s-ukai@eng.hokudai.ac.jp

1. Introduction

The conservation of nature and fossil fuel resources has become a global concern. Improvements in the thermal efficiency of power plants promote the economic utilization of fossil fuels and thereby reduce CO_2 emissions, which is very significant in view of the global agreements on provisions to counter the greenhouse problem. The ultra-supercritical (USC) and Advanced USC conditions significantly increase the thermal efficiency of coal-firing steam turbines; further improvements in efficiency can be achieved by utilizing gas turbines that combust gasified coal or liquid natural gas. This chapter introduces fundamental metallurgical principles to develop high temperature materials, role of microstructure and strengthening mechanism. Ferritic and austenitic heat-resistant steels, Ni-base superalloys as well as Fe-base and Ni-base ODS alloys are included, and their temperature ranges for applicability are finally summarized.

2. Heat-Resistant Ferritic Steel

2.1 *Alloy Design*

(a) Microstructure and alloying design

The loss of creep rupture strength in heat-resistant ferritic steels has been extensively investigated for a number of 9–12Cr steels. Figure 1 represents schematic diagram of microstructure on the tempered martensite, which consists of packet, block and lath boundaries inside prior-austenite grain.[1] $M_{23}C_6$ carbide and MX nitrides are precipitated at those boundaries. The proposed mechanisms for the loss of creep rupture strength are ascribed to the occurrence of microstructure degradation during creep exposure through (i) coarsening and/or dissolution of fine MX carbonitride precipitates and new undesirable phase precipitates and (ii) preferential recovery of microstructures and dislocations in the vicinity of prior-austenite grain boundaries. With regard to (i), fine precipitates contribute

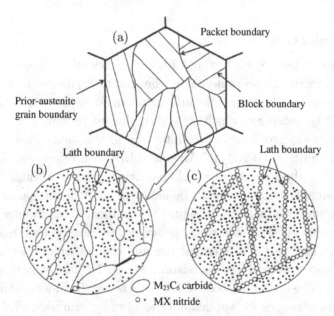

Fig. 1. Schematic diagram of (a) the typical microstructure of the tempered martensite, (b) conventional steels and (c) developmental steels containing MX nitrides in heat-resistant ferritic steels.[1]

to high creep resistance by the pinning action of the dislocation movement. These MX-type fine precipitates are coarsened and lose the dislocation pinning during creep. From the current results of microstructure observations, fine MX precipitates are dissolved by the precipitate of a new phase such as the Z-phase, which is a complex nitride of the form Cr(Nb,V)N; this consumes existing MX precipitates, and thus, the creep strength is lost after longer durations. With regard to (ii), the preferential recovery of the microstructure is induced at the vicinity of the prior-austenite grain boundaries and is accompanied by precipitate coarsening due to the enhanced diffusion through the grain boundaries. This microstructure change promotes the onset of acceleration creep and hence causes premature ruptures. In order to improve the creep rupture strength of 9–12Cr ferritic steels, it is necessary to restrict the dislocation recovery within the lath-martensite structure and stabilize the fine precipitates. The onset of acceleration creep can be retarded by stabilizing the lath and block microstructures and by suppressing local deformation at the grain boundaries by preventing the dissolution and coarsening of precipitates. General concept of alloying design of ferritic heat-resistant steels is shown in Fig. 2. Detail is shown in the following sections.

(b) Precipitation strengthening

The addition of small amounts of V and Nb significantly stabilizes precipitates such as V_4C_3 and NbC at high temperatures. Based

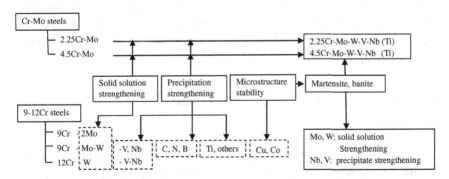

Fig. 2.   General concept of alloying design for ferritic heat-resistant steels.

on the creep rupture test at 650°C for 10,000 hours, the optimum content of V is around 0.20%. Since NbC is difficult to be dissolved even at high temperature, the addition of Nb content must be restricted. If all Nb-carbides are dissolved into a matrix at the normalizing temperature of 1050°C, grain growth occurs and the notch toughness reduces. The addition of 0.05% Nb is recommended, where a small amount of NbC remains during the normalizing heat treatment. The 9–12Cr ferritic steels are strengthened by very fine V_4C_3 and NbC, which are precipitated at tempering heat treatment. Moreover, subsequent carbide precipitates such as $M_{23}C_6$ strengthen during the creep test. Therefore, the equation, $V/51 + Nb/93 < C/12$, must be satisfied for the formation of $M_{23}C_6$ carbide along with V_4C_3 and NbC precipitates[2,3] where V, Nb, C indicate the mass % of each element. Fujita and Asakura concluded that the peak strength at high temperatures occurred at 0.6–0.7 atomic % for $(V+Nb)/C$. The excess carbon is consumed with the formation of $M_{23}C_6$ carbide.

(c) Solution hardening

Mo and W are typical solution hardening elements. A balance of Mo and W contents gives a trade-off between creep rupture strength and toughness. Since W-addition suppresses coagulation and the growth in carbides ($M_{23}C_6$ and M_6C), the addition of more W with less Mo in 12Cr ferritic steels tends to strengthen the steels at high temperatures, whereas too much W leads to loss of ductility/toughness. An optimum value of Mo-equivalent, that is Mo + 1/2W (mass %), is given as 1.5 for 12Cr ferritic steels.[4] The composition of 2 mass % W and 0.5 mass % Mo is usually selected for an optimal balance between creep rupture strength and ductility/toughness.

2.2 Materials for Fossil Power Plant

(a) Boiler tubes

Figure 3 shows the progress of the development in heat-resistant ferritic steels for boiler tubes.[5] The target of the creep strength for developing heat-resistant steels is usually standardized as 100 MPa

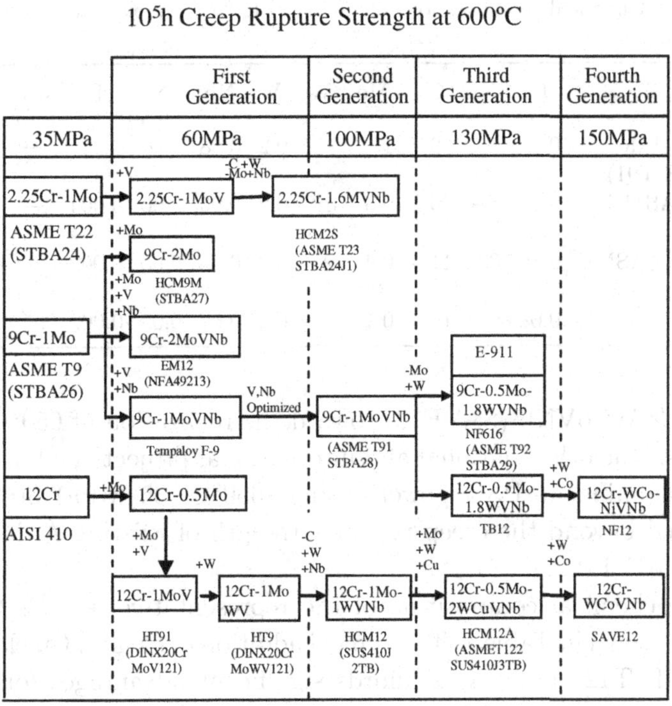

10^5h Creep Rupture Strength at 600°C

	First Generation	Second Generation	Third Generation	Fourth Generation
35MPa	60MPa	100MPa	130MPa	150MPa

2.25Cr-1Mo —+V→ 2.25Cr-1MoV —-C+W -Mo+Nb→ 2.25Cr-1.6MVNb

ASME T22 (STBA24) —+Mo→ 9Cr-2Mo   HCM2S (ASME T23 STBA24J1)

+Mo HCM9M (STBA27) +V +Nb

9Cr-1Mo —→ 9Cr-2MoVNb   E-911

ASME T9 (STBA26) +V +Nb EM12 (NFA49213)   -Mo +W   9Cr-0.5Mo-1.8WVNb

9Cr-1MoVNb   V,Nb Optimized   9Cr-1MoVNb   NF616 (ASME T92 STBA29)

Tempaloy F-9   (ASME T91 STBA28)   12Cr-0.5Mo-1.8WVNb   +W +Co   12Cr-WCo-NiVNb

12Cr —+Mo→ 12Cr-0.5Mo   TB12   NF12

AISI 410   +Mo +V   -C +W +Nb   +Mo +W +Cu   +W +Co

+W   12Cr-1MoV → 12Cr-1MoWV → 12Cr-1Mo-1WVNb → 12Cr-0.5Mo-2WCuVNb → 12Cr-WCoVNb

HT91 (DINX20Cr MoV121)   HT9 (DINX20Cr MoWV121)   HCM12 (SUS410J 2TB)   HCM12A (ASMET122 SUS410J3TB)   SAVE12

Fig. 3. Development progress of 2.25-12Cr ferritic steels for boiler tubes.[5]

for 100,000 hours (service duration) at the operation temperature. Heat-resistant steels are classified as 2.25Cr, 9Cr, and 12Cr ferritic steels, of which microstructures are composed of tempered martensite containing a small amount of delta ferrite.

Regarding 9Cr ferritic steels, Mod.9Cr-1Mo (ASME T91) was developed by the US Oak Ridge National Laboratory, and an ultra-supercritical (USC) power plant was realized for the first time at an operation temperature of 593°C (1100°F) by employing ASME T91 steel. 9Cr-0.5Mo-1.8W-VNb (ASME T92), designated as NF616, was developed by replacing some amount Mo with W by Nippon Steel Corp. A similar strategy led to the joint development of 12Cr-0.5Mo-2W-CuVNb (ASME T122), designated as HCM12A, by Mitsubishi Heavy Industries Ltd., and Sumitomo Metal Industries Ltd. Further development is underway for 12Cr-WCoNiVNb (NF12)

Table 1: Chemical composition of 9–12Cr ferritic steels for boiler tubes (mass %).

Steels	C	Ni	Cr	Mo	W	V	Nb	N	B	Co	Cu	Fe
Mod.9Cr-1Mo (ASME T91)	0.10	—	9	1	—	0.2	0.08	—	—	—	—	Bal
NF616 (ASME T92)	0.07	—	9	0.5	1.8	0.2	0.05	0.06	0.004	—	—	Bal
HCM12A (ASME T122)	0.10	0.3	11.5	0.4	2.0	0.2	0.05	0.05	0.002	—	0.8	Bal
NF12	0.08	0.5	11	0.15	2.6	0.2	0.07	0.05	0.002	2.5	—	Bal

and 12Cr-WCoVNb (SAVE12) toward the realization of 650°C class of steels. The other national and international projects in Japan and Europe aiming at USC power plants (650°C, 35 MPa) are being promoted beyond the creep rupture strength of existing ASME T92 and ASME T122.

The chemical composition of the representative 9–12Cr ferritic steels is shown in Table 1. The typical addition of Co and Cu elements in ASME T122 and NF12 affords significant advantages for stabilizing the austenite without lowering A_1 point. These features not only restrict the delta-ferrite formation that leads to a degradation of toughness at the welding parts but also increase the tempering temperature. The addition of a small amount of B affects the strengthening of ferritic steels due to the enhancement of martensitic transformation and the stabilization of carbide precipitates.

(b) Turbine rotor

Turbine rotor steels contain more carbon and are tempered at a lower temperature since they are required to have superior tensile properties at relatively lower temperatures. An alloying design similar to that of boiler tubes was adopted for turbine rotor steels, except for the higher C content and lower tempering temperature. In the 1950s, the 1CrMoV steel rotor was used for the main steam temperature of 538°C. In the 1970s, the steam temperature was increased to 566°C by using GE-10.5Cr1MoVNbN steel. On the basis of TAF steel (Tokyo–Akutagawa–Fujita), which was originally

Table 2:   Chemical composition and heat-treatment condition of turbine rotor steels (mass %).

Steels	C	Ni	Cr	Mo	W	V	Nb	N	B	Co	Fe	Main steam tempera- ture (°C)
GE	0.10	0.60	10.5	1.0	—	0.20	0.085	0.06	—	—	Bal	566
TR1100	0.14	0.60	10.2	1.5	—	0.17	0.055	0.04	—	—	Bal	593
HR1100	0.15	0.64	10.2	1.2	0.34	0.15	0.05	0.05	—	—	Bal	600
TR1200	0.13	0.80	11.0	0.15	2.5	0.20	0.06	0.05	—	—	Bal	600–630
HR1200	0.09	0.51	11.0	0.23	2.66	0.22	0.07	0.02	0.018	2.53	Bal	630–650

developed by Fujita in 1956, TR1100 was developed for a class of 593°C rotors and TR1200 for 600°C, by means of which the USC plant was commercialized.

New 12%Cr rotor steels, HR1100 (12CrMoWVNb) and HR1200 (12CrWMoCoVNbB), were designed by Hitachi Ltd., and a significant increase in creep rupture strength was achieved by the addition of W for HR1100 and Co and B for HR1200. HR1100 rotor steel has already been used in USC commercial operation at 600°C, and HR1200 will be the first attempt to achieve an operation temperature of 630–650°C. The chemical composition of these steels is summarized in Table 2.

2.3 *Advanced Ferritic Steels*

R&D has been conducted at the National Institute of Materials Science (NIMS) on advanced ferritic steels for applications to the main steam pipe and header of a USC power plant at 650°C and 35 MPa. An attempt was made to stabilize the lath martensitic microstructure in the vicinity of prior-austenite grain boundaries by the addition of B to reduce the coarsening rate of $M_{23}C_6$ carbides. The chemical composition and heat-treatment conditions of 0.0139B steel (139 ppm B added) are listed in Table 3.[6] However, $M_{23}C_6$ and MC carbides are coarsened at prolonged period of exposure at elevated temperature, but nitrides are more stable since precipitate coarsening according to Ostwald ripening is enhanced by solubility

Table 3:   Chemical composition and heat-treatment condition of the advanced ferritic steels developed by NIMS (mass %).[6]

Steels	C	Ni	Cr	Mo	W	V	Nb	N	B	Co	Fe	Norm/Temp
0.0139B steel	0.078	—	8.99	—	2.91	0.19	0.050	0.003	0.0139	3.01	Bal	1080°C × 1 h/ 800°C × 1 h
0.002C steel	0.002	—	9.12	—	2.96	0.20	0.060	0.049	0.0070	3.09	Bal	1100°C × 0.5 h/ 800°C × 1 h

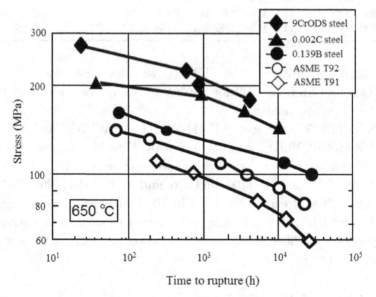

Fig. 4.   Creep rupture data for the advanced ferritic steels (0.002C steel and 0.139B steel) together with ASME T91 and T92 and ODS steel at 650°C.[6]

of solute element in the matrix. Thermo-Calc analysis indicates that solubility of Cr, V, Nb in the Fe matrix in equilibrium with carbide is extremely higher than that in equilibrium with nitride. Therefore, an attempt was also made to stabilize the microstructure by using a dispersion of MX nitrides instead of $M_{23}C_6$ and MC carbides. The 0.002C steel shown in Table 3 was developed on the basis of Fe-9Cr-3W-3Co-0.2V-0.05Nb including 0.05N and less carbon.

The creep rupture data for the 0.0139B and 0.002C steels at 650°C are shown in Fig. 4 along with those of T91, T92, and 9CrODS steel.[6] The creep rupture strength of 0.0139B steel is significantly higher than that of T91 and T92. The strength level of 0.002C steel is significantly greater than that of developmental steels. This is because a large number of V- and Nb-nitride precipitates having sizes of less than 10 nm are distributed along the prior-austenite grain boundaries as well as along the lath, block, and packet boundaries after tempering. These microstructures cause a significant decrease in the inter-particle distance along boundaries and hence induce a significant increase in the pinning force for boundary migration. The highest creep rupture strength in ODS steel is also plotted in this figure. The solubility of solute element in Fe matrix in equilibrium with oxide particles is extremely reduced far beyond that with nitrides; this is a major reason why nano-sized oxide particles are so stable and 9CrODS exhibits higher strength. A detail of the microstructure control and dispersion strength in ODS steels is described in Section 5.1.

3. Heat-Resistant Austenitic Steels

3.1 *Alloy Design*

(a) General view

The austenitic steels have critical advantages such as low diffusion coefficient and large solubility of constituent elements in FCC crystal structure, which make it possible to design for superior high temperature strength and corrosion resistance to the ferritic steels. Figure 5 shows the concept of alloy design for heat-resistant austenitic steels to improve creep strength through the modification of existing steels.[7] The chemical composition can be classified into four types, and precipitation strengthening by $M_{23}C_6$, TiC, and NbC and the solution strengthening by Mo and W applied basically in the same way as the heat-resistant ferritic steels.

Three types of austenitic steels are derived from 18%Cr-8%Ni steels (Type 304): Type 316 solution-strengthen by Mo, Type 321 precipitation-strengthen by TiC and Type 347 precipitation-strengthen by NbC. These austenitic steels were originally developed

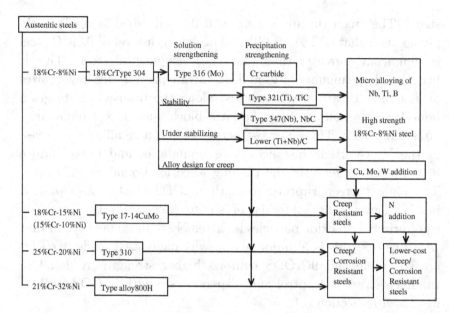

Fig. 5.   General concept of alloy design for austenitic heat-resistant steels.[7]

for the materials of chemical equipment, placing emphasis on corrosion resistance. Accordingly, further enhancement of creep strength was conducted by means of precipitation strengthening. Cu addition significantly improves the creep strength by coherent Cu phase precipitate in austenitic matrix. 15%Cr–15%Ni, 25%Cr–20%Ni, 21%Cr–32%Ni steels with a full austenitic phase structure are capable of high creep strength, although they are costly due to high Ni content. Steels containing 20%Cr and more Cr are likely to have excellent oxidation resistance, and a costly Ni content of at least 30% is required to maintain a full austenitic structure. Low-cost, high-strength, highly corrosion-resistant austenitic steel can be designed by adding about 0.2%N to reduce the Ni content.

(b) Precipitation strengthening

TiC and NbC are used for the precipitation strengthening in austenitic heat-resistant steels, and amounts of adding Ti and Nb atoms are reduced against the C atoms. The excess C forms Cr carbide. Figure 6 shows the peak of the creep rupture strength in

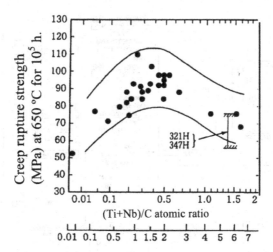

Fig. 6. Effect of (Ti + Nb)/C ratio on creep rupture strength of 18Cr10NiNbTi steel.[8]

18Cr10NiNbTi steel against the ratio of (Ti + Nb)/C in atomic ratio, indicating that Ti and Nb contents relative to the C content can be optimized at around 0.4 in atomic ratio.[8]

The Cu addition substantially enhances the creep strength through finely dispersed Cu-rich particle precipitates. This effect is effective for Cu addition in 3%. When Cu addition exceeds 3%, the strength tends to be saturated and creep rupture ductility declines.[9]

3.2 *Materials for Heat Exchanger and Boiler Tubes*

The development progress of austenitic boiler steels is presented in Fig. 7, and their chemical compositions are given in Table 4. For the grade of 18%Cr–8%Ni, various improvements have been made to improve the creep strength in order to apply to the superheater and reheater of boiler tubes. The steam oxidation resistance is also improved at their inner surface through finer grains. AISI 304H, AISI 321H, AISI 347H, and AISI 316H are still used for fossil-fired power plants operating under conventional steam condition. AISI 347H has the highest allowable stress among the four types, and it was improved to make a fine-grained structure for steam oxidation

S. Ukai

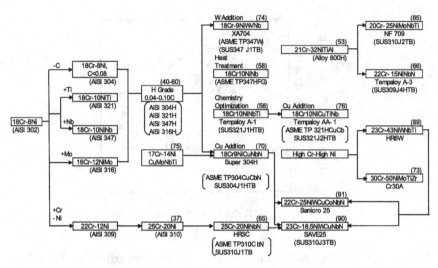

Fig. 7.   Development progress of austenitic boiler steels[10]

resistance and creep strengthening, designed as ASME 347HFG. This steel is widely employed in the superheater tubes of the USC pressurized power plants operated up to 600°C.

Newly developed 20–25%Cr austenitic steels have excellent high temperature strength as well as being relatively inexpensive. They are practically applied as high temperature strength and corrosion resistant steels. Allowable stresses for HR3C,[11] NF709,[12] and Tempaloy A-313[13] are far higher than that of Alloy 800H. SAVE25[14] and Sanicro25[15] were also developed by using 0.25% addition of N to stabilize the austenitic structure. HR6W is 23%Cr–43%Ni and strengthen in by addition of 6% W. SAVE25, Sanicro25 and HR6W exhibit the highest creep rupture strength of around 90 MPa at 700°C and 100,000 h. Realizing 700°C class USC pressurized power plant requires 100 MPa (70 MPa in allowable stress) at 100,000 h for creep rupture strength. Therefore, the creep rupture strength of all of 18%Cr–8%Ni steels and 20%–25%Cr austenitic steels is insufficient, and their applicability would be approximately limited to 660°C and 680°C. Indeed, austenitic steels such as Discaloy (13%Cr–25%Ni–3%MoAlTi) and A286 (15%Cr–26%Ni–1.5%MoAlTi) have

Table 4: Nominal chemical composition of austenitic steels for boilers.[10]

	Steels		C	Si	Mn	Ni	Cr	Mo	W	V	Nb	Ti	B	Others
18%Cr–8%Ni	TP304H	18Cr–8Ni	0.08	0.6	1.6	8.0	18.0	—	—	—	—	—	—	—
	TP316H	16Cr–12NiMo	0.08	0.6	1.6	12.0	16.0	2.5	—	—	—	—	—	—
	TP321H	18Cr–10NiTi	0.08	0.6	1.6	10.0	18.0	—	—	—	—	0.50	—	—
	TP347H	18Cr–10NiNb	0.08	0.6	1.6	10.0	18.0	—	—	—	0.80	—	—	—
	TP347HFG	18Cr–10NiNb(FG)	0.08	0.6	1.6	10.0	18.0	—	—	—	0.80	—	—	—
	Tempaloy A-1	18Cr–10NiNbTi	0.12	0.6	1.6	10.0	18.0	—	—	—	0.10	0.08	—	—
	Super304H	18Cr–9NiCuNbN	0.10	0.2	0.8	9.0	18.0	—	—	—	0.40	—	—	3.0Cu, 0.1N
15%Cr–15%Ni	XA704	18Cr–9NiWVNb	0.03	0.3	1.5	9.0	18.0	—	2.5	0.3	0.3	—	—	0.2N
	Tempaloy AA-1	18Cr–10NiCuTiNb	0.10	0.3	1.5	10.0	18.0	—	—	—	0.3	0.2	0.02	3.0Cu, 0.1N
	17-14CuMo	17Cr–14NiCuMoNbTi	0.12	0.5	1.7	14.0	16.0	2.0	—	—	0.40	0.30	0.008	3.0Cu, 0.1N
	15-15N	15Cr–15NiMoWNbN	0.12	0.7	1.5	15.0	15.0	1.5	1.5	—	1.0	—	—	0.1N
	AN31	15Cr–13NiMoNbN	0.10	0.5	1.5	14.0	16.0	1.5	—	0.5	1.0	—	—	0.1N
	Esshete1250	15Cr–10Ni6MnVNbTi	0.12	0.5	0.6	10.0	15.0	1.0	—	0.2	1.00	0.06	—	—
	12R72	15Cr–10NiMoNbVB	0.10	0.4	2.0	15.0	15.0	1.0	—	—	—	0.3	0.006	—
20%–25%Cr	TP310	25Cr–20Ni	0.08	0.6	1.6	20.0	25.0	—	—	—	—	—	—	—
	HR3C	25Cr–20NiNbN	0.08	0.4	1.2	20.0	25.0	—	—	—	0.45	—	—	0.2N
	Alloy 800H	21Cr–32NiTiAl	0.08	0.5	1.2	32.0	21.0	—	—	—	—	0.50	—	0.4Al
	Tempaloy A-3	22Cr–15NiNbN	0.05	0.4	1.5	15.0	22.0	—	—	—	0.70	—	0.002	0.15N
	NF709	20Cr–25NiMoNbTi	0.15	0.5	1.0	25.0	20.0	1.5	—	—	0.20	0.10	—	—
	SAVE25	22.5Cr–18.5NiWCuNbN	0.10	0.1	1.0	18.0	23.0	—	1.5	—	0.45	—	—	3.0Cu, 0.2N
High Cr– High Ni	Sanicro25	22Cr–25NiWCuNbN	0.08	0.2	0.5	25.0	22.0	—	3.0	—	0.3	—	—	3.0Cu, 0.2N
	CR30A	30Cr–50NiMoTiZr	0.06	0.3	0.2	50.0	30.0	2.0	—	—	—	0.20	—	0.03Zr
	HR6W	23Cr–43NiWNbTi	0.08	0.4	1.2	43.0	23.0	—	6.0	—	0.18	0.08	0.003	—

Chemical composition (mass%)

been demonstrated for 650°C operation in Wakamatsu, Japan.

3.3 *Advanced Austenite, Fe–Ni, and Ni-based Alloys*

Temperature over 700°C would require Fe–Ni based alloys with improved creep strength or Ni-based alloys such as Inco617 (Ni–22%Cr–13%Co–9%MoAlTi) or Waspaloy (Ni–19%Cr–14%Co–4.5%MoAlTi). From the perspective of practical application, fabricability would need to be improved to enable the manufacture of the large components.

Figure 8 shows the relationship between 100,000 h creep rupture strength and temperature for various materials. Along with consideration of Inco617, efforts are also being made worldwide to improve Inco706 (Ni–16%Cr–36%Fe–3%NbAlTi) and to develop alloys modifying the low thermal expansion Inver-type Ni alloy. Of these, LTES700[16] (Ni–12%Cr–18%MoAlTi) was developed, having the low thermal expansion coefficient similar to ferritic steels. Furthermore, FENIX-700 (Ni–16%Cr–36%Fe–2%NbAlTi)[17] modified Inco 706 and TOS-1X (Ni–23%Cr–10%Mo–15%CoAlTiB)[18] have been developed aiming at 700°C class steam turbine rotor application.

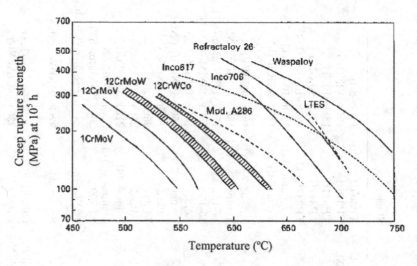

Fig. 8. Comparison of 10^5 h creep rupture strength of turbine rotor alloys.

4. Ni-based Superalloys

4.1 *Progress for Gas Turbine and Jet Engine*

When significant resistance to loading under tensile, fatigue, and creep conditions is required, the Ni-based superalloys have emerged as the preferred materials for operation temperatures greater than approximately 800°C. This is the case for the gas turbines used for jet propulsion and electricity generation. Over the latter part of the twentieth century, alloy and process developments have enabled dramatic improvements in the performance of superalloys. Figure 9 provides a perspective for the alloy and process developments that have occurred since the first superalloys emerged in the 1940s.[19] The introduction of improved casting methods, followed by the introduction of processing by directional solidification, enabled significant improvements in the creep rupture strength; this was due to the formation of columnar microstructures in which the transverse grain boundaries were removed. The grain boundaries

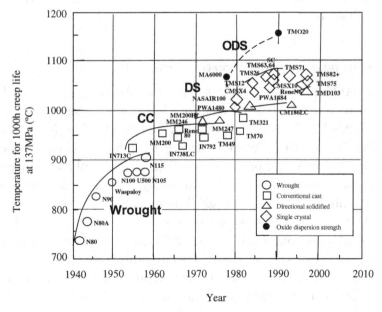

Fig. 9.   Evolution of the high temperature capability in the Ni-based superalloys since 1940s.[19]

are completely removed in single-crystal (SC) superalloys. This allowed the removal of grain boundary strengthening elements such as B and C that were added in the casting in directionally solidified superalloys; this enables better heat treatment to reduce the micro-segregation and eutectic content induced by casting. Nowadays, SC superalloys are being used in gas turbine engines, although the castings in the columnar and equiaxed forms are still used in many cases.

Table 5 lists the compositions of the representative SC Ni-base superalloys with γ' precipitation strengthening.[19] First-generation superalloys such as PWA1480 and TMS-26 contain substantial quantities of γ' hardening elements such as Al, Ti, and Ta, and grain-boundary-strengthening elements such as C and B are absent. Second-generation alloys such as PWA1484, Rene'N5, CMSX-4, and TMS-82 are characterized by a 3% Re concentration. The Re concentration increased to approximately 6% in third-generation alloys such as Rene'N6, CMSX-10, and TMS-75. Fourth-generation SC superalloys such as MC-NG and TMS-138 are characterized by the addition of Ru.

4.2 *Composition and Crystal Structure*

The Ni–Al binary system exhibits a number of solid phases other than the disordered Ni(Al) solid solution. There are ordered structures such as Ni_3Al, Ni_5Al_3, NiAl, Ni_2Al_3, and $NiAl_3$. The Ni(Al) solid solution is an FCC crystal structure with lattice sites randomly occupied by either Ni or Al atoms. The Ni_3Al-γ' phase is of great importance considering the role that it plays in conferring strength to the superalloys. The γ' phase exhibits a cubic $L1_2$ crystal structure with Al atoms at the cube corners and Ni atoms at the face centers. The β-NiAl compound exhibits a CsCl-type BCC crystal structure.

All the Ni-based superalloys listed in Table 5 are strengthened by the Ni_3Al-γ' precipitate. These commercial alloys often incorporate a large number of alloying elements, most of which are taken from the d-block of transition metals in the periodic table.[20] A first class of elements includes V, Cr, Fe, Co, Y, Mo, Ru, Hf, W, Re, and

Table 5: The chemical composition of SC Ni-based superalloys as a representative (mass %).[19]

Alloy	Co	Cr	Mo	W	Al	Ti	Nb	Ta	Hf	Re	C	B	Others	Ni	Generation
PWA1480	5	10	—	4	5	1.5	—	12	—	—	—	—	—	Bal	1st
CMSX-2	4.6	8	0.6	8	5.6	1	—	9	—	—	—	—	—	Bal	1st
TMS-26	8.2	5.6	1.9	10.9	5.1	—	—	7.7	—	—	—	—	—	Bal	1st
PWA1484	10	5	2	6	5.6	—	—	9	—	3	—	—	—	Bal	2nd
Rene'N5	8	7	2	5	6.2	—	—	7	0.2	3	—	—	—	Bal	2nd
CMSX-4	9	6.5	0.6	6	5.6	1	—	6.5	0.1	3	—	—	—	Bal	2nd
TMS-82+	7.8	4.9	1.9	8.7	5.3	0.5	—	6	0.1	2.4	—	—	—	Bal	2nd
YH61	1	7.1	0.8	8.8	5.1	—	0.8	8.9	0.25	1.4	0.07	0.02	—	Bal	2nd
Rene'N6	12.5	4.2	1.4	6	5.75	—	—	7.2	0.15	5.4	0.05	0.004	0.01Y	Bal	3rd
CMSX-10	3	2	0.4	5	5.7	0.2	0.1	8	0.03	6	—	—	—	Bal	3rd
TMS-7	12	3	2	6	6	—	—	6	0.1	5	—	—	—	Bal	3rd
MC-NG	<0.2	4	1	5	6	0.5	—	5	0.1	4	—	—	4Ru	Bal	4th
TMS-138	6	3	3	6	6	—	—	6	0.1	5	—	—	2Ru	Bal	4th
TMS-162	5.8	2.9	3.9	5.8	5.8	—	—	5.8	0.1	4.9	—	—	6Ru	Bal	4th

Ir; these prefer to partition to the austenite γ. The atomic radii of these elements are similar to that of Ni. The second class of elements comprise the Ni_3Al-γ' precipitate. These elements include Ti, Nb, and Ta. B, C, and Zr constitute a third class of elements that tend to segregate to grain boundaries. These elements have very different atomic sizes as compared to that of Ni, which induces the formation of carbides and borides.

4.3 Mechanical Properties

Ni-based superalloys have relatively high yield and ultimate tensile strengths, with the former often ranging from 900 MPa to 1300 MPa and the latter ranging from 1200 MPa to 1600 MPa at room temperature. Figure 10 shows typical data of yield stress versus temperature for SC superalloys.[20] The strengthening arises from solid-solution, grain-size strengthening, and interaction of dislocations with Ni_3Al-γ' precipitates. It should be noted that the Ni-based superalloys exhibit a remarkable characteristic, in that the yield stress increases with temperature, typically until approximately 800°C, although most alloy systems exhibit decreased deformation stress with increasing temperature. At temperatures greater than 800°C, the yield stress rapidly decreases and does not maintain resistance at 1200°C. This anomalous yield stress behavior, namely, positive temperature dependence, is explained as follows. Upon deformation,

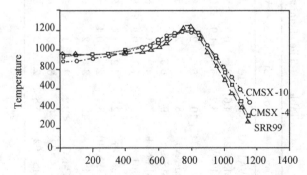

Fig. 10.   Yield stress versus temperature in SC Ni-based superalloys.[20]

the applied stress induces the cross-slip of segments of the γ' superpartial dislocations from the {111} slip plane to the cross-slip plane {001} due to the lower antiphase boundary energy of the latter as compared to that of the former; this is a unique process in L1$_2$-type intermetallic compounds. The cross-slip dislocations lock the moving dislocations in the {111} plane. The strength increases with temperature due to the formation of more locking dislocations with easy thermal activation. This process is known as the Kear–Wisdorf mechanism.[20]

Another characteristic feature of Ni-based superalloys arises from the γ/γ' interface coherency. SC Ni-based superalloys possess a typical two-phase γ/γ' microstructure in the cuboidal form, as shown in Fig. 11.[21] The γ/γ' interface remains coherent and the interfacial energy remains low if the lattice misfit δ between the lattice parameters of the γ and γ' phases is not too large. Therefore, the properties of superalloys inherently depend on the coherency of the γ/γ' interfaces. Small values of δ are preferred. With the application of an external stress, the discrete cuboidal γ' precipitates coalescence into rafts or rods that are aligned perpendicular or parallel to the applied stress direction. Since most commercial directionally solidified and SC alloys exhibit a negative misfit and are used to sustain tensile loads, rafts are generally formed perpendicular to the applied stress direction. This situation is also shown in Fig.

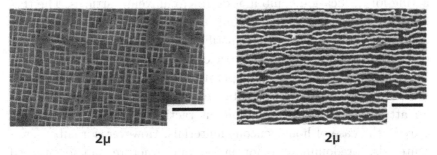

2μ   2μ

Fig. 11. The γ/γ' microstructure in SC Ni-based superalloys.[21] (a) cuboidal morphology at manufacturing; (b) directional coarsening resulted from formation of γ' rafts aligned perpendicular to the applied stress direction after creep damage.

11. Once rafting is complete, microstructural damage can accumulate rapidly and the tertiary creep rates are accelerated.

5. ODS Alloys

5.1 *Fe-based ODS Steels*

(a) General views

Incorporating fine stable oxide dispersoid particles in an alloy matrix can significantly improve the high temperature capabilities of materials. Such oxide dispersion strengthened (ODS) alloys generally retain useful strength up to a relatively high fraction of their melting points where other strengthening mechanisms, e.g., precipitation hardening or solid solution strengthening, rapidly lose their effectiveness. These oxide particles can be introduced by the mechanical alloy processes invented by Benjamin, followed by consolidation and thermomechanical processing.

The creep behavior of ODS alloys can be characterized by a threshold stress beyond which plastic deformation occurs. Several theoretical and experimental studies have been devoted to interpret the deformation behavior of ODS alloys, particularly focusing on the origin of threshold stress for deformation. It was observed that the threshold stress becomes lower than the so-called Orowan stress; in contrast, the threshold stress is significantly higher than that acting for dislocation climb motion to circumvent particles. The theoretical analysis by Srolovitz confirmed that an attractive particle-dislocation interaction can be justified at high temperatures.[22] The dislocation detachment from particles can act as the threshold stress for deformation. The high temperature creep behavior of Fe-based and Ni-based superalloys can be reasonably predicted in terms of an attractive particle-dislocation interaction model.[23] This is true only in the case of homogeneous materials. However, for substantial materials, predominant deformation and fracture should proceed through deformations in weak regions such as the grain boundaries.

For Fe-based ODS alloys, high-Cr ferritic steels are utilized as a matrix with finely distributed yttria particles that are several to several tens of nanometers in size, which considerably improves the high temperature strength. Two groups of ODS ferritic steels have been commercialized or are under development. The first is a 20%Cr-ODS ferritic steel containing 5% Al and it exhibits superior resistance to oxidation and corrosion in hot gases at temperatures greater than 1000°C; tubes, sheets, and bars made from this steel are commercially used in various kinds of stationary and high-temperature components in turbines, combustion chambers, diesel engines, and burners. The second is used for the fuel cladding of nuclear fast reactors due to its superior resistance to radiation resistance; further, it exhibits excellent creep strength and dimensional stability at a temperature of 700°C under severe neutron exposure.

The basic chemical compositions of the representative ODS ferritic steels are shown in Table 6. There are similar compositions of 20%Cr ODS ferritic steels containing 0.5% Y_2O_3 for the high-temperature component. The 9–14% Cr ODS ferritic steels containing Mo/W were developed for the FBR fuel cladding. Furthermore, it is necessary to modify the manufacturing process to overcome the difficulties in tube formation and ductility-loss due to χ phase formation under irradiation.

(b) Nanostructure control

It has been found that a fine distribution of Y_2O_3 particles, which is essential for improving the high temperature strength of ODS ferritic steels, can be attained in the process of the dissolution

Table 6: Basic chemical composition of ODS ferritic steels including Japanese FBR fuel cladding (mass %).

Steels	Cr	Mo	W	Ti	Al	Dispersoid	Fe	Application
Incoloy MA956	20	—	—	0.50	4.5	$0.5Y_2O_3$	Bal	turbine, combustion
PM2000	19	—	—	0.50	5.5	$0.5Y_2O_3$	Bal	
Incoloy MA957	14	0.3	—	1	—	$0.25Y_2O_3$	Bal	FBR fuel
DT2203Y05	13	1.5	—	2.2	—	$0.5Y_2O_3$	Bal	
DT2906	13	1.5	—	2.9	—	Ti_2O_3, TiO_2	Bal	
JAEA-ODS	9	—	2.0	0.2	—	$0.35Y_2O_3$	Bal	

of oxide particles during MA processing. The thermodynamically
stable Y_2O_3 particles are forcibly decomposed into the ferritic steel
matrix during the MA process, and subsequent annealing induces
the precipitation of fine oxide particles at elevated temperature
greater than $1000°$C. It was also discovered that the co-addition of Ti
during MA processing promotes the decomposition of Y_2O_3 and then
induces the precipitation of Y-Ti-O complex oxide with an ultra-fine
size during the annealing heat treatment.[24] The microstructure of
nano-oxide particles as obtained by transmission electron microscopy
(TEM) is shown in Fig. 12.

The crystalline structure of nano-oxide particles has been studied by means of high-resolution transmission electron microscopy
(HRTEM); cubic $Y_2Ti_2O_7$ (Y:Ti = 1:1) is stable when there is a
substantial supply of oxygen to the reaction, whereas insufficient
supply of oxygen leads to the restricted formation of the hexagonal
structure of Y_2TiO_5 (Y:Ti = 2:1).[25] With regard to the oxide
particle/matrix interfacial coherency, Karlsruhe's group proposed
the following crystalline relationship between $Y_2Ti_2O_7$ and ferritic
matrix,[26] which corresponds to the so-called Kurdjumov–Sachs rela-
tion: $(1\bar{1}\bar{1})_{oxide}//(1\bar{1}0)_{ferrite}$ and $[110]_{oxide}//[111]_{ferrite}$. This result

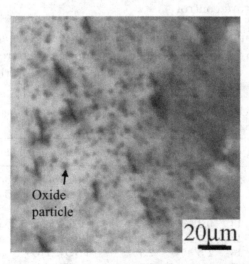

Fig. 12. Typical microstructures of oxide particles in 9Cr-ODS ferritic steel by
means of TEM.

suggests that nano-oxide particles precipitate from the ferritic matrix while maintaining crystalline coherency or partial coherency with the ferritic matrix such that they decrease the free energy in the system from the extremely higher energy state induced by mechanical alloying. Based on these results, 0.2%Ti is added to the basic composition of 9Cr-ODS ferritic steel as 9Cr-0.13C-0.2Ti-2W-0.35Y$_2$O$_3$ (mass%).

(c) Creep rupture property

The ODS ferritic steels have an advantage from the viewpoint of creep strength when they are used at temperature around 1000°C. However, in the second series of ODS ferritic steels listed in Table 6, excellent radiation resistance was noticed for FBR fuel cladding application, although utilization temperature is relatively low around 700°C. The creep rupture strengths of the manufactured 9Cr–ODS ferritic cladding at 650°C, 700°C and 750°C are shown in Fig. 13 in comparison with those of typical fast reactor cladding, namely, HT9 (ferrite) and PNC316 (austenite).[24] These curves were predicted based on the Larson–Miller–Parameter (LMP) method. The creep strength of 9Cr–ODS ferritic steel is significantly greater than that of HT9, and is also superior to that of PNC316 beyond 1000 h at

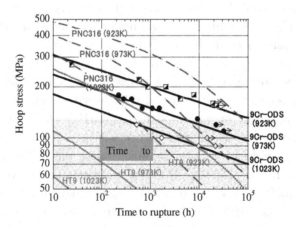

Fig. 13.   Creep rupture curves of 9Cr-ODS ferritic steel cladding at temperatures of 650°C, 700°C, and 750°C, compared with those of HT9 and PNC316.[24]

750°C. It should be noted that stress acting on the fuel cladding ranges below 120 MPa in the FBR condition; the lower stress and longer times are dominant regions in fuel pin system. As compared to advanced ferritic steels such as B-added steels (0.0139B) and fine nitride precipitate steels (0.002°C) shown in Fig. 4, 9Cr-ODS ferritic steel has a slight superiority at 650°C; furthermore at temperatures above 700°C, the effectiveness of oxide dispersion strengthening becomes more obvious. This noticeably improved strength in 9Cr-ODS ferric claddings is attributable to the extremely fine distribution of oxide particles. The theoretical analysis of the threshold stress for deformation using the dispersion parameters suggested that a dispersion strengthening level determined in terms of the attractive particle-dislocation interaction model should be greater than the measured strength level; the deformation and fracture could arise from weakened area such as grain boundaries. It is noticed that the creep rupture strength of 9Cr-ODS steel cladding can potentially be improved further. In fact, recent developments have demonstrated that further improvements in creep rupture strength can be successfully realized by controlling the packet boundaries in the thermomechanical treatment process.

(d) Oxidation and corrosion resistance

Incoloy MA956 and PM2000 listed in Table 6 are based on the conventional Kanthal and Fecralloy heating conductor alloys. The oxidation resistance of the ODS ferrite is mainly determined by the aluminum content of the alloys. The total aluminum content usually varies between 4 wt% and 5 wt%. In combustion and hot-gas atmospheres, these iron-based ODS ferrites form excellent and protective alumina or alumina-chromium(III) oxide spine layers. These oxide layers which are highly resistant to thermodynamic influences provide protection against oxidation and corrosion up to temperatures of 1350–1400°C. In addition to the chemical and thermal stability, these alumina layers have extremely good adhesive strength on the ODS ferrite. As shown in Fig. 14, the excellent oxidation resistance of the MA956 is provided during isothermal heat treatment when compared with Ni-based superalloys.[27]

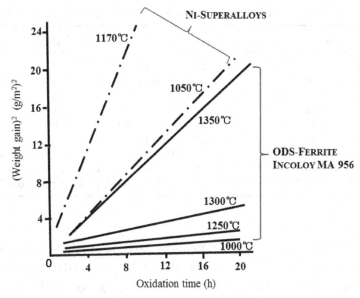

Fig. 14.   Comparison of oxidation weight gain versus duration for MA956 and Ni-based superalloy.[27]

5.2 *Ni-based ODS Alloys*

(a) General views

Benjamin, who originally developed the MA process, made a technical breakthrough in the dispersion of oxide particles into Ni-based superalloys. The utilization temperature of conventional Ni-based superalloys is limited to 1000°C for stable conditions of the γ' phase. In order to extend the utilization temperature beyond 1000°C, various types of Ni-based ODS superalloys have been developed. Table 7 lists the commercial ODS superalloys. Y_2O_3 oxide is widely used as a dispersoid since it is similar to Fe-based ODS alloys. However, it has recently been suggested that the added Y_2O_3 oxide particles are chemically changed to several types of Y–Al–O, e.g., $Y_3Al_5O_{12}$(YAG), $Y_4Al_2O_9$(YAM), and $YAlO_3$(YAP), in the process of the dissolution of Y_2O_3 oxide particles during the mechanical alloying and subsequent Y–Al–O complex oxide precipitation during annealing. It is noted that this behavior is exactly the same as

Table 7:   Chemical composition of Ni-based ODS superalloys (mass %).[27]

Alloy	Al	Cr	Ti	Ta	W	Mo	Fe	Zr	C	B	Y_2O_3	others	Ni
MA6000	4.5	15	2.5	2	4.0	2.0	—	0.15	0.05	0.01	1.1		Bal
MA760	6.0	20	—	—	3.5	2.0	—	0.15	0.05	0.01	0.95		Bal
MA754	0.3	20	0.5	—	—	—	1.0	—	0.05	—	0.6		Bal
MA758	0.3	30	0.5	—	—	—	1.0	—	0.05	—	0.6		Bal
MA757	4.0	16	0.5	—	—	—	—	—	0.05	—	0.6		Bal
PM3030	6.0	17	—	2	3.5	2.0	—	0.15	—	—	0.9		Bal
PM1000	0.3	20	0.5	—	—	—	3.0	—	—	—	0.6		Bal
TMO-2	4.2	5.9	0.8	4.7	12.4	2.0	—	0.05	0.05	0.01	1.1	Co:9.8	Bal
TMO-20	5.5	4.3	1.1	6.0	11.6	1.5	—	0.05	0.05	0.01	1.1	Co:8.7	Bal

that exhibited by Fe-based ODS alloys, except for chemical forms of Y–Al–O other than Y–Ti–O. The almost Ni-based ODS superalloys, listed in Table 7, are strengthened not only by oxide particles but also by γ' (Ni_3Al) in the γ phase. MA series alloys were manufactured by Inco Alloys International Inc., (Special Metals Inc., at present) and PM is manufactured by Plansee Inc. The NIMS in Japan produced TMO-2 and TMO-20 alloys. Inconel MA6000 was first developed for applications to advanced aerospace gas turbines, while MA760 was developed for industrial applications that require extremely high resistance to high temperature corrosion and oxidation.

(b) Creep rupture properties

The effects of oxide particles on the creep strength of the Ni-based ODS superalloys have been investigated by several authors. The threshold stress concept, arising from the interaction between oxide particles and dislocations, has become widely accepted. This is true for the Ni-based ODS superalloys having recrystallized coarser grains. Very fine grains ($1\,\mu$m), which are produced by the as-consolidated condition, exhibit no threshold stress and higher strain rate at lower loading stress. This fact is attributed to the occurrence of grain-boundary sliding in a manner similar to 9Cr-ODS ferritic steels. Therefore, Ni-based ODS superalloys are

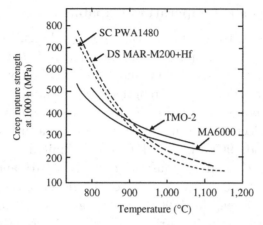

Fig. 15.   Creep rupture strength at 1000 h versus temperatures for Ni-based ODS superalloys (MA6000 and TMO-2), compared with conventional directionally solidified and SC superalloys.[28]

used after recrystallization heat-treatment, which produces coarser grains (1 mm) and highly elongated grains with high aspect ratios. The creep rupture strength of these Ni-based ODS superalloys are compared with conventional superalloys without Y_2O_3 in Fig. 15.[28] TMO-2 and MA6000 are strengthened by combining γ'-phase precipitation hardening at intermediate temperature and oxide dispersion strengthening at high temperature, where γ' precipitation hardening is attributed to dislocation interaction with the coherent γ/γ' interfaces in a manner similar to Ni-based superalloys. It is worth noting that TMO-2 and MA6000 have superior creep rupture strength than conventional Ni-based superalloys, particularly at higher temperatures greater than 1000°C.

With regard to the industrial applications of Ni-based ODS superalloys, MA754 is used as the stationary blade of an aerospace gas turbine, etc. On the other hand, SC type Ni-based superalloys are being applied to rotary blades. Considering the superior high-temperature strength of Ni-based ODS superalloys at temperatures greater than 1000°C, their industrial applications are expected to increase with further improvements in the high temperature strength as well as the ductility and formability.

6. Summary for Temperature Capability

In this chapter, the metallurgical alloying design, strengthening mechanism and development for the heat-resistant ferritic and austenitic steels, Ni-based superalloys and ODS alloys were briefly reviewed. High temperature strengths of representative alloys in each category are represented in Fig. 16, where creep rupture strengths at 1000 h are selected in terms of service temperature. Applicability of ferritic Mod. 9Cr-1Mo steel (T91) is limited to temperature less than 600°C, and the austenitic steels (304H and 800H) are less than 700°C. By applying the oxide dispersion strengthening, creep rupture strength of 9CrODS is significantly improved over austenitic steels even in ferritic steel matrix, however its service temperature is restricted to A_{C1} temperature (870°C) due to α/γ

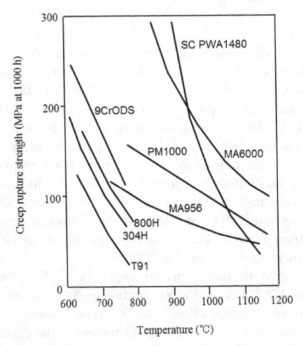

Fig. 16. Temperature capability for the creep rupture strength at 1000 h. Ferritic heat-resistant steel, T91: less than 600°C Austenitic heat-resistant steels, 304H 800H: less than 700°C Ni-based superalloy, SC PWA1480: less than 1000°C Ni-based ODS superalloy, PM1000 and MA600: less than 1200°C Fe-based ODS steel, MA956: less than 1300°C.

transformation. On the other hand, the full ferritic steel MA956 without phase transformation maintains adequate creep strength at the temperature beyond 900°C, and provides excellent resistance for the high temperature oxidation up to 1300°C due to surface Al_2O_3 formation, as shown in Fig. 14. Concerning Ni-based alloys, the SC PWA 1480 in the first generation represents extremely high strength at the temperature below 1000°C, but the creep rupture strength is abruptly reduced by over 1000°C due to γ' dissolution into γ matrix. The creep rupture strength of Ni-based ODS superalloy PM1000 is higher than MA956 at the same temperature region in accordance with a limited diffusion coefficient in FCC type of PM1000 as compared with BCC type of MA956. Ni-based MA6000 strengthened by not only dispersed oxide particles but also γ' precipitate has the highest creep rupture strength at the temperatures above 1000°C.

We must take into account the other factors, oxidation and corrosion resistance as well as the melting temperature range, which restrict the applicable temperature. The solidus and liquidus temperatures of MA6000 range from 1280°C–1340°C; thus, this property restricts the application of MA6000 to below 1200°C. MA956 was derived from the conventional Kanthal alloy, and has a high melting point of 1480°C as well as excellent oxidation resistance shown in Fig. 14. These properties permit higher temperature application of 1300°C. MA956 is interesting and particularly suitable for stationary and high temperature component in turbine and diesel engine construction.

References

1. *Comprehensive Nuclear Materials*, Vol. 4, R.J.M. Konings *et al.*, (Eds.) (Elsevier), p. 108.
2. T. Fujita, K. Asakura, T. Sawada and Y. Otoguro, *Metall. Trans. A* **12** (1981), p. 1071.
3. T. Fujita, *Netsushori* **27** (1987), p. 4. (in Japanese)
4. T. Fujita, *The Thermal and Nuclear Power* **42**(11) (1991), p. 1485.
5. F. Masuyama, *Advanced Heat Resistant Steels for Power Generation*, R. Viswanathan and J.W. Nutting (Eds.) (IOM Communications Ltd., London, 1999), p. 33.

6. F. Abe, *Current Opinion in Solid State and Materials Science* **8** (2004), p. 305.
7. F. Masuyama, Transaction of Iron Steel Institute of Japan, *ISIJ International* **41** (2001), p. 612.
8. T. Shinoda and R. Tanaka, *Bulletin Japan Institute of Metal* **11** (1972), p. 180.
9. Y. Sawaragi, K. Ogawa *et al.*, *Sumitomo Metals* **48** (1992), p. 50.
10. F. Masuyama, 8th *Intl. Con. Materials for Advanced Power Engineering, Liege*, Juelich GmbH, (Forschungszentrum, Juelich, 2006).
11. Y. Sawaragi, H. Teranishi *et al.*, *Sumitomo Metals* **37** (1985), p. 166.
12. M. Kikuti, M. Sakakibara *et al.*, *Int. Con. High Temperature Alloys*, (Petten, Netherlands, 1985).
13. A. Toyama, Y. Minami *et al.*, *CAMP-ISIJ* **1** (1988), p. 928.
14. H. Semba, M. Igarashi *et al.*, *Proc. Inter. Conf. Power Engineering* **97**(2), JSME, Tokyo, 1997.
15. R. Rautio and S. Bruce, Advances in materials technology for fossil power plants, (*ASM International*, Metals Park, Ohio, 2005), p. 274.
16. R. Yamamoto, Y. Kadoya *et al.*, in Advances in materials technology for fossil power plants, (*ASM International*, Metals Park, Ohio, 2005), p. 623.
17. H. Imano *et al.*, in Advances in materials technology for fossil power plants, (*ASM International*, Metals Park, Ohio, 2005), p. 575.
18. K. Imai, private communication.
19. Harada and T. Yokokawa, *Materia Japan* **42**(9), (2003), p. 621. (in Japanese).
20. R.C. Reed, *The Superalloys — Fundamentals and Applications* (Cambridge University Press, 2006).
21. T.M. Pollock and S. Tin, *J. Propulsion and Power* **22**(2), (2006), p. 361.
22. D.J. Srolovits, R.A. Petkovic-Lution and M.J. Luton, *Acta. Metall.* **31**(12), (1983), p. 2151.
23. E. Arzt, *Res. Mechanica* 31, (1991), p. 399.
24. S. Ukai and M. Fujiwara, *J. Nucl. Mater.* **307–311**, (2002), p. 749.
25. S. Ohtsuka, S. Ukai, M. Fujiwara, T. Kaito and T. Narita, *J. Nucl. Mater.* **329–333**, (2004), p. 372.
26. M. Klimiankou, R. Lindau and A. Moslang, *J. Nucl. Mater.* **329–333**, (2004), p. 347.
27. H.D. Hedrich, *Proc. Conf. Mechanical Alloying*, 1990, pp. 217–230.
28. R.F. Singer and G.H. Gessinger, *Powder Metallurgy of Superalloys*, G.H. Gessinger (Ed.) (Butterworth & Co., London, 1984), p. 213.

Chapter 4

High Temperature Corrosion Problems
in Refineries, Chemical Process Industries
and Petrochemical Plants

Pasi Kangas

Sandvik Materials Technology
Mumbai Pune Road, Dapodi, Pune 411012, India

1. Introduction

In chemical process industries, the choice of material is important and sometimes complex. In high temperature processes, there is often a combination of properties required by the material of construction such as high temperature strength, fatigue resistance, creep resistance, formability and corrosion resistance. Metallic materials in general are unstable in high temperature applications, but when there are conditions so that an oxide layer may be formed, sufficient protection can be achieved. In general, the demand is that there is a sufficiently high oxygen activity to form a stable oxide layer, which is mainly achieved by the combination of alloying elements in the material.

2. High Temperature Corrosion

High temperature corrosion research is about the art of identifying the mechanisms of corrosion and the conditions under which the surface layer can be protective. If the surface oxide scale is not protective for some reason, spalling may take place as shown in Fig. 1.

P. Kangas

Fig. 1.   Heat treated SAF 2507 at 1050°C/24 h in open air atmosphere.

Fig. 2.   Oxidation of APMT and Fe-35Ni–25Cr material at 1100°C for 2300 hours in an oxygen atmosphere.

The example is taken from heat treatment of the duplex stainless steel SAF 2507 at 1050°C whereby MoO_3 has been formed, so that the formation of a protective oxide layer is hindered.

Protective oxides in high temperature materials are predominantly, chromia- alumina- or silica oxides. An example of a protective surface layer of aluminum oxide, Al_2O_3, formed on the surface of APMT[1] aluminum oxide layer may protect against carburization or coke formation. Figure 2 shows APMT and alloy 800 exposed to a oxygen atmosphere for 2300 hours at 1100°C. APMT is very little affected by the exposure whereas for a Fe–35Ni–25Cr material, severe oxidation has occurred.[2]

3. High Temperature Materials

Most high temperature materials are austenitic as creep strength is required in addition to corrosion resistance. One exception within Sandvik is the ferritic alloy 4C54, which has good corrosion resistance owing to its high chromium content. Other ferritic alloys are APMT, a FeCrAl material containing 3% molybdenum.

Austenitic high temperature alloys such as 304H, 316H, 321H, and 347H find use in applications where the temperature is below 850°C. Carburization is a typical problem at higher temperatures.[5] For resistance to carburization and sulfidation, type 310H material may be used at temperatures up to about 1100°C. For temperatures up to 1150°C, the alloys 253MA, 353MA, Sanicro 31HT, and Sanicro 70, are suitable. The alloys 253MA and 353MA are alloyed with silicon and cerium for improved corrosion resistance and thereby become resistant to carburization and nitriding owing to the formed oxide scale. 253MA is typically used in applications where 304H and 316H have insufficient resistance to oxidation or carburization. Example applications for 253MA are thermocouple protection tubes and muffle tubes for wire annealing. 353MA is used in more demanding petrochemical applications like steam crackers where carburization is a problem. The higher nickel content in 353MA makes it a suitable choice in cracked ammonia gas due to its resistance to nitriding.

Sanicro 31HT has a combination of corrosion resistance and creep strength, especially in nitriding atmosphere, and is used for pigtails and headers in plants for production of synthetic gas for ammonia and for muffle tubes in wire annealing.

Almost all steels can become embrittled due to formation of brittle precipitations in the material at high temperatures. Sigma-phase is a brittle intermetallic precipitation formed in the temperature range 600–850°C. The sigma-phase precipitation is very much dependent on the composition of the material. Chromium

Table 1: Overview of high temperature alloys and their suitability in various environments. 0 reference, – Not suitable, + (+ + +) Suitable to very suitable.[8]

Type	Steel grade	Sulfidation	Oxidation	Carburization	Nitridation	HT fluegas
Ferritic	4C54/TP446-1	++	+	0	0	++
	APM/APMT	N/A	+++	+++	N/A	N/A
Austenitic	TP 304H	0	0	0	0	–
	TP 316H	0	0	0	0	–
	253MA	+	++	++	++	++
	TP310S/H	+	+	+	+	+
High Nickel	Alloy 800H	–	–	–	–	–
	Alloy 800HT	0	0	++	++	0
	353MA	+	++	++	++	++
	Alloy 825	–	–	–	–	–
Ni base	Alloy 601	–	–	–	–	–
	Alloy 600	0	++	++	++	+

and molybdenum are elements that promote sigma-phase formation whereas elements like nickel and nitrogen slow down the process of sigma-phase formation. Sandvik 253 MA is alloyed with nitrogen and is less prone to sigma-phase formation compared to 310H.

Ferritic steels alloyed with higher amounts of chromium are very sensitive to sigma-phase formation and they are also prone to embrittlement caused by spinodal decomposition in the temperature range 400–550°C (475°C-embrittlement). If a material is heat treated at about 1100°C, most embrittling phases are dissolved and the material becomes ductile.

4. Some Typical Applications and Case Stories for High Temperature Materials

4.1 *Petrochemical Plants*

Acetic acid (or sometimes Acetone) is cracked to form ketene and water using a 0.2–0.3% triethyl phosphate catalyst at 700–750°C. Reaction takes place in heating coils.

Acetic acid and ketene are reacted to form acetic anhydride following stabilization of the ketene using small quantities of ammonia.

In this process, coke forms in the preheater and cracker tubes when the acetic acid is cracked and de-coking is carried out using a mixture of steam and air.

The higher material temperature during de-coking can cause the catalyst to decompose to form phosphorous, which may lead to corrosion of the cracker tube and preheater.

Most commonly used grades are Si-alloyed Sicromal 12 (24Cr, 1.5Al, 1.2Si) and AISI 310. Both grades have a problem with sigma-phase embrittlement in this temperature range.

The solution is to use a more structure stable material with better resistance to sigma-phase precipitation such as 253MA or Sanicro 31HT.

In EDC-crackers, materials can suffer from severe grain boundary corrosion due to the high concentration of HCl and Cl_2 in the reaction gas. If the tubes suffer from premature failure, there is a high probability that the material delivered has had initial grain boundary attacks from the pickling during production. In some cases, external fins are soldered to the tube OD to increase the heat transferring area. The solder should not contain lead (Pb) if a Ni-base alloy is used in the tube material.

4.2 *Ethylene Furnace Tubes*

Ethylene cracking furnace is an application, where furnace tubes are exposed to hydrocarbons at a skin temperature of about 1050°C, which puts high demands on the material, both from the corrosion and the strength point of view. Ethylene furnace tubes are subjected to a carburizing, and coking environment at 1000–1100°C, when ethylene is cracked from the gaseous or liquid hydrocarbon feed stock by steam to ethylene. 353 MA mostly is the best solution due to its resistance to carburization and oxidation as well as its high creep strength. In certain cases, 253 MA or Sanicro 31HT can be chosen. APMT, with its Al-oxide, has been proven to resist coke formation very well and is therefore an even better alternative.

4.3 *Lance Tubes in Steel Plant Blast Furnaces*

In a steel manufacturers blast furnace, the ASTM 310 stainless steel lance tubes have a relatively short life span, caused by fast outer surface oxidation and high inner erosion. With the aim of increasing lance tube service life, the company decided to find a more heat-resistant tube material. The R&D department ran trials with tubes in AISI 310 grade and 253 MA, due to its very good resistance to isothermal and cyclic oxidation, as well as maintaining structural stability at high temperatures. After promising test results, the new tubes in 253 MA were installed in the main plant. The changeover to 253 MA has increased the service life of the lance tubes by 100%, resulting in fewer maintenance stops and increased productivity. By further fine-tuning the lance installation, powder mixtures and injection pressure, the manufacturer expects to increase lance tube life even further.

4.4 *Muffle Tubes in Heat Treatment Furnaces*

Muffle furnaces are most often used in wire drawing mills and in bundy tube production, but they can also be found in other applications such as razor blade production and tube annealing. They are used to shield a product from the environment of the furnace during heat treatment, and to create conditions for a more even temperature distribution.

In most cases, some protective gas is fed into the muffle tube. This shielding gas can be hydrogen, nitrogen, cracked ammonia or endogas ($CO + H_2$). Some of these gases are very aggressive and will shorten the life of the muffle tubes significantly. In some annealing furnaces, the temperature can reach above 1200°C, but temperatures between 800°C and 1120°C are most common.

These high temperatures often result in a short service life, leading to frequent maintenance stops for muffle tube replacements. Cracked ammonia is the most aggressive environment in this application. It causes rapid nitriding of the tube material, which leads to a loss of the mechanical strength. By selecting an alloy with higher nickel content, the service life can be extended. In this environment,

Fig. 3.   Muffle tubes.[6]

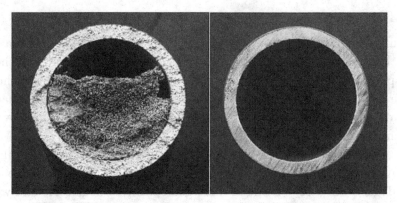

Alloy 800HT after 6 months   **Kanthal APM after 24 months**

Fig. 4.   Alloy 800HT and Kanthal APM tubes after use as muffle tubes in a wire drawing plant.[2]

Alloy 601 offers much longer service life. Alloy 800HT as a suitable material where pure nitrogen or a gas mixture of nitrogen and hydrogen is used. Nitrogen is a less severe environment than cracked ammonia.

The endogas will cause a rapid carburization, which also reduces the muffle tube's mechanical strength. In these conditions, 353 MA is the most cost-effective material. For severely carburizing conditions, Kanthal APM or APMT is a better choice than 353 MA. Hydrogen is a less aggressive environment. In these conditions, the most cost-effective material is 253 MA, followed by Alloy 800HT. 4C54 is a cost-effective choice for annealing of carbon steel as it is done in a lower temperature range.

An automobile manufacturer used muffle tubes in a bundy tube application where they experienced coking problems. The material used was alloy 800HT and by switching to 353MA, the service life was doubled and coking problems were reduced significantly.

Common failures in muffle tubes

(a) Brittle fractures

If the fracture occurs outside the furnace at the gas exit end, this may be due to zinc-induced liquid metal embrittlement (LME). This can happen if the producer uses zinc-phosphate as a bonder between the wire material and the lubricant in the drawing process. Solutions to the problem could be to use another bonder or a low nickel-alloy (e.g., 253MA).

If the fracture occurs inside the furnace, this is most probably due to the atmosphere inside the tube. It can be due to nitriding (e.g., the protective gas is nitrogen, a mixture of nitrogen + hydrogen, or cracked ammonia) or carburization due to insufficient decreasing before annealing of the wire. A sudden failure has been noted when degreasing routines have changed, so that degreasing is insufficient. Embrittlement due to nitriding and carburization may result in a change of the magnetic properties of the materials, whereby austenitic materials may become magnetic. By checking the austenitic stainless steel tubes with an ordinary magnet, it may reveal if the affected area of the tube is very local, i.e., only a short length of the tube is magnetic, which indicates that the material may have been locally over-heated and that the problem also is connected to the furnace heating device.

(b) Elongation and rupture

Elongation and rupture is connected to the creep properties of the tube material. This type of failure can occur when the furnace is frequently cycled up and down in temperature and using a dead weight as a constant load to keep the tubes straight. Check the weight used to keep the tubes straight. The solution can be to use a stronger tube material (e.g., 353MA) or to unload some of the weight used, at least at working temperature and during the cooling of the furnace.

(c) Corrosion through the wall

When a muffle tube suffers from corrosion through the entire wall there could be a number of different causes. The most common cause is the corrosion caused by aggressive elements from the ID. Some such elements are Sodium (Na), Potassium (K), and Chlorine (Cl_2), in some cases also Sulfur (S) can be active. The source of these elements is mainly the soap used as lubricant and chemicals used in the decreasing process.

In a few cases, the corrosion has been caused by electrical heating elements that have sagged from the roof and created a short circuit over the tubes.

4.5 *Recuperator Tubes*

Recuperators save energy by using the waste heat to preheat process gas, reducing fuel consumption and costs. They are found in many energy intensive industries, such as steel industry, glass production and carbon black production. As energy prices increase, recuperators may reduce costs significantly, even in industries with low energy consumption. A current trend is to increase the heat recovery by increasing air preheating temperatures, resulting in more energy savings.

Depending on process parameters, such as flue gas temperature, flue gas composition and heat recovery ambitions, the material in the recuperator must be carefully selected.

253 MA has been used in many different types of recuperators for many years. The success of this alloy is due to its excellent oxide

Fig. 5.   Radiant recuperator.[7]

properties that prevent material degradation, especially in carbon black and direct reduction of iron ore (DRI) applications.

When changing from an air preheat of 650°C using common ferritic or austenitic materials to an air preheat of 800°C using the high performance 253 MA, considerable savings can be expected.

4C54 is optimal in the glass industry with flue gas temperatures of 1,300°C and higher because this material combines corrosion resistance with low thermal expansion.

Kanthal APM and Kanthal APMT are high performance materials for use at the highest temperatures since they form a protective layer of Aluminum Oxide.

A steel mill tripled the lifetime of their air preheaters by switching from TP310H to 253MA in the recuperator. Fewer maintenance shutdowns resulted in higher productivity in the hot rolling mill, as well as lower operating costs thanks to a longer service life.

Common failures

(a) Cracking

The cracking is usually caused by using the material in a temperature range where brittle phases are formed, e.g., sigma phase. Try to verify the working temperature of recuperator (note that the working temperature may deviate from the design temperature).

(b) Bowing and deformation of front tubes

This is usually caused by using a poor design, e.g., the design does not allow thermal expansion of the tubes. In some cases, this can be a result of over-heating due to reduced or blocked air flow.

(C) Corrosion

In the recuperator application, corrosion occurs mainly on the front tubes, i.e., the first tubes that meet the flue gas. Here the composition in the fuel plays a very important role. Some trace elements important for this type of failure are sodium, potassium, sulfur, vanadium and chlorine. If the corrosion occurs behind the front tubes, then elements that form liquid oxides (e.g., vanadium and molybdenum) can be important. In some cases in carbon rich atmospheres, for instance in the DRI application, the tube material may suffer from 'metal dusting' and then it could be of interest to discuss materials like Sanicro 60X9 (Alloy 601).

References

1. Johanna Nockert Olovsjö, Dilip Chandrasekaran and Fernando Rave, Performance of alumina forming alloy Kanthal Apmt® in aggressive environments, *Oxidation of Metals*.
2. *Kanthal PM Tubes, Ardiant Tubes and Heating Systems*, Kanthal Printed matter 6-B-2-3.
3. *Stainless steels for high temperature applications*, Sandvik Printed matter, S-130-ENG, 02-2008.
4. Sales Training Manual, Sandvik Materials Technology, 1998.
5. P. Szakalos, J. Lindblom and L. Lindé, Carburizing of chromia and alumina forming stainless steels at 750°C, in *14th Int. Corrosion Congress* (ICC), Cape Town, South Africa, Sept., 1999.
6. Sandvik Stainless Muffle Tubes, Sandvik Printed matter S1301-ENG, 02-1999.
7. Sandvik Stainless High Temperature Grades. Sandvik Printed MAtter S-130-ENG, 06-1998.
8. The Sandvik Handbook to Physical Metallurgy S-GE038B, Sandvik Publication 2013.

Chapter 5

High Temperature Corrosion Problems in Coal-based Thermal Power Plants

A.S. Khanna
Department of Metallurgical Engineering and Materials Science
Indian Institute of Technology Bombay
Mumbai 400076, India

1. Introduction

In a thermal power plant, steam is formed by heating water using a fuel, which rotates the turbine blades and thus drives a generator to produce electricity. The steam is condensed after it passes through the turbine and recycles through condenser as per the Rankine cycle, shown in Fig. 1 along with the schematic of the power plant. The first practical electricity generating system, using a steam turbine, was designed and made by Charles Parsons in 1885 and was used for lighting an exhibition in Newcastle.

There is variation in the design of thermal power plants due to the use of different fuel resources, to heat the water. These resources can be fossil fuels, which can be further classified as coal, lignite, oil, bio-waste etc. The other set of resources are nuclear, hydel, wind and solar. Out of all these, the fossil fueled thermal power plants are well known and widely used and perhaps are the largest source of power production in several countries. However, they produce enormous CO_2 emissions to the atmosphere, and are also the cause of corrosion to the materials used for power plant construction. Thus, efforts are being made to reduce the carbon dioxide emissions and also work on better methods to reduce corrosion of materials in the power plant.

A.S. Khanna

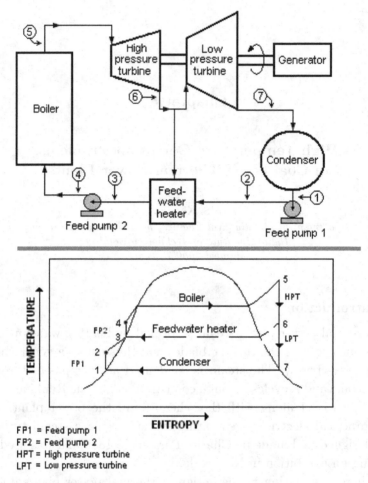

FP1 = Feed pump 1
FP2 = Feed pump 2
HPT = High pressure turbine
LPT = Low pressure turbine

Fig. 1. Schematic of a thermal power plant with Rankine cycle.[1]

The demand for energy is increasing enormously every day due to better standard of living, increased population and increasing requirements of infrastructure. Currently, demand for all global energy is increasing at an average rate of approximately 2% per annum.[2] This rate is expected to continue or may increase even further. The energy efficiency of a conventional thermal power plant, in terms of energy produced as a percent of the heating value of the fuel consumed, is typically 33–48%.[1] As with all heat engines, this efficiency is limited, and governed by the laws of thermodynamics. In comparison to coal-based power plants, hydel power plants are about 90% efficient in converting the energy of falling water into electricity.

According to the Carnot cycle, higher efficiencies can be attained by increasing the temperature of the steam. Sub-critical fossil fuel power plants can achieve 36–40% efficiency. Super critical designs have efficiencies in the range of 40%, with new "ultra critical" designs using pressures around 30 MPa and multiple stages reheat, reaching about 48% efficiency. Currently, many nuclear reactors operate at efficiency not greater than 30–32%, at temperatures and pressure lower than that of coal-based power plants. This is mainly because of the size of the reactor, which limits the pressure that can be reached, since the pressurized vessel is very large and contains the entire bundle of nuclear fuel rods. Some advanced reactors, such as the very high temperature gas cooled nuclear reactor, and super-critical water reactor, would operate at temperatures and pressures similar to current coal plants, producing comparable thermodynamic efficiency.

2. Indian Scenario of Power Generation

India has the world's fifth largest electricity generating capacity. According to the Ministry of Power, the total installed capacity of power in India is 181,558 MW. India has abundant sources of power production. Thermal power in India accounts for roughly two-thirds of the power generated in India which includes gas, liquid fuel and coal. Reserves for thermal power generation include 59 billion tons of mineable coal, 775 million metric tons of oil reserves and natural gas reserves of 1,074 billion cubic meters. Other prominent and fast-growing sources of power are hydro, wind, solar, nuclear, biomass and industrial waste, etc. Presently, out of the total power produced, coal-based is 54.8%, 9.75% is gas based and 0.66% is oil based, hydro contributes for 21% of power, while nuclear production is 2.63% and the rest 11.1% is collectively produced by renewable energy sources such as small hydro project, biomass gas fire, urban and industrial waste power and wind energy.[3]

3. Types of Thermal Power Plants

Thermal power plants can be divided based on the type of combustion or gasification: boilers, internal reciprocating engines, and
combustion turbines. In addition, combined cycle and cogeneration
systems increase efficiency by utilizing heat lost by conventional
combustion systems. The type of system is chosen based on the loads,
the availability of fuels, and the energy requirements of the electric
power generation facility. Other ancillary processes, such as coal
processing and pollution control, must also be performed to support
the generation of electricity. The following subsections describe each
system and then discuss ancillary processes at the facility (USEPA
1997).

3.1 *Steam Boiler*

As stated above, the conventional steam-producing thermal power
plants generate electricity through a series of energy conversion
stages: fuel is burned in boilers to convert water to high pressure
steam, which is then used to drive a steam turbine to generate
electricity. Heat for the system is usually provided by the combustion
of coal and other types of waste or recovered fuel. High temperature,
high pressure steam generated in the boiler then enters the steam
turbine. At the other end of the steam turbine is the condenser, which
is maintained at a low temperature and pressure. Steam rushing from
the high pressure boiler to the low pressure condenser drives the
turbine blades, which powers the electric generator. Low-pressure
steam exiting the turbine is finally sent back to boiler through
condenser shell/tubes which are maintained at a low temperature
by the flow of cooling water.

Coal and lignite are the most common fuels in thermal power
plants, which are designed to use pulverized coal or crushed coal.
Several types of coal-fired steam generators are in use, however, they
are classified based on the characteristics of burning the coal. In
fluidized-bed combustors, fuel materials are forced by gas into a state
of buoyancy. The gas cushion between the solids allows the particles
to move freely, thus flowing like a liquid.

3.2 Combustion Turbines

Gas turbine systems operate in a manner similar to steam turbine systems except that combustion gases are used to turn the turbine blades instead of steam. In addition to the electric generator, the turbine also drives a rotating compressor to pressurize the air, which is then mixed with either gas or liquid fuel in a combustion chamber. The greater the compression, the higher the temperature and thus the higher the efficiency of gas turbine. Higher temperatures, however, generate more SO_2 and NO_x. Exhaust gases are emitted to the atmosphere from the turbine. Unlike a steam turbine system, gas turbine systems do not have boilers or a steam supply, condensers, or a waste heat disposal system. Therefore, capital costs are much lower for a gas turbine system than for a steam system.

3.3 Combined Cycle

Combined cycle generation is a configuration using both gas turbines and steam generators. In a combined cycle gas turbine (CCGT), the hot exhaust gases of a gas turbine are used to provide all, or a portion of, the heat source for the boiler, which produces steam for the steam generator turbine. This combination increases the thermal efficiency to approximately 50–60%. Combined cycle system may have multiple gas turbines driving one steam turbine. Combined cycle systems with diesel engines and steam generators are also sometimes used.

4. Corrosion in Coal-Based Power Plants

Coal is a natural occurring material and has therefore composition, varying from place to place. If the coal was pure, there would be any corrosion problem. Burning of coal would lead to the following oxidation reaction:

$$C(s) + O_2(g) = CO_2(g). \tag{1}$$

The energy liberated would be used to heat boiler water. The main reason for corrosion in coal-based power plants is the inherent

Table 1:   List of impurities in coal along with their caloric values in different coal types.[4]

English Designation	Volatiles (%)	C Carbon (%)	H Hydrogen (%)	O Oxygen (%)	S Sulfur (%)	Heat content (kJ/kg)
Lignite (brown coal)	45–65	60–75	6.0–5.8	34–17	0.5–3	<28,470
Flame coal	40–45	75–82	6.0–5.8	>9.8	~1	<32,870
Gas flame coal	35–40	82–85	5.8–5.6	9.8–7.3	~1	<33,910
Gas coal	28–35	85–87.5	5.6–5.0	7.3–4.5	~1	<34,960
Fat coal	19–28	87.5–89.5	5.0–4.5	4.5–3.2	~1	<35,380
Forge coal	14–19	89.5–90.5	4.5–4.0	3.2–2.8	~1	<35,380
Non-baking coal	10–14	90.5–91.5	4.0–3.75	2.8–3.5	~1	35,380
Anthracite	7–12	>91.5	<3.75	<2.5	~1	<35,300

impurities in coal. Coal is a complex and relatively dirty fuel that contains varying amounts of sulfur and a substantial fraction of non-combustible mineral constituents, commonly called ash. Distribution of various impurities in different types of coals along with their caloric values is shown in Table 1.

Let us now look into the effect of these impurities on corrosion:

(a) Sulfur is the impurity which is responsible for causing sulfidation, oxidation and hot corrosion of boiler tubes. Parts per million (ppm) level impurity of S in coal can form Sulfur dioxide, (SO_2) and Sulfur trioxide (SO_3) when coal is burnt. The amount of various sulfur products released depends upon the air to fuel ratio as shown in Fig. 2.[3] Even in reducing conditions, during low power requirements, when oxygen supply is reduced, "S" can react with hydrogen, formed due to reduction of water, resulting in H_2S formation which reacts with steel directly, much faster than even oxygen or sulfur dioxide or trioxide.

(b) The second most prevalent impurity in coal is salt, usually NaCl and KCl. These salts can react with SO_2/SO_3 to form sodium sulfate, which deposits on the boiler tubes along with ash and can lead to hot corrosion when melt at certain partial pressures of

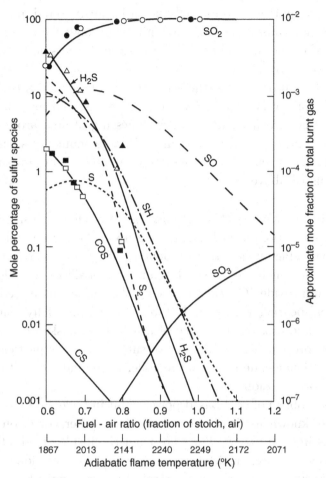

Fig. 2.   Concentration of various sulfur based species, released during burning of coal as a function of air/fuel ratio.[5]

SO_2/SO_3 that leads to the formation of complex salts.[6,7] These complex salts melt at the temperature of the boiler operation.

(c) The sulfur transfer from the environment to the alloy is determined by direct reaction of the oxide-forming alloying elements with the SO_2.[8-10] Strong oxide formers, such as chromium and aluminum, tend to not only form protective oxide layers but also increase the driving force for sulfidation.[8]

(d) Silica impurity is also a well-known impurity found in coal. This does not cause corrosion, but hits the boiler tubes along with ash after the coal is burnt and causes severe erosion to the boiler tubes.

(e) The fourth problem is the ash formation tendency of coal. Many coals form very high ash and others less ash. Formation of large quantity of ash results in severe messing of area around boiler tubes and can lead to erosion with large chunks of silica or hot corrosion when ash contains large amount of sodium sulfate as discussed above.

In an industrial context, fly ash usually refers to ash produced during combustion of coal. Depending upon the source and makeup of the coal being burned, the components of fly ash vary considerably, but all fly ash includes substantial amounts of silicon dioxide (SiO_2) and calcium oxide (CaO), both being the important ingredients in many coal-bearing rock strata. Two main effects of fly ash are on corrosion and erosion. Presence of ash in fact masks or dilutes the corrosion process, while presence of silica and other particles cause erosion. When ash deposits on the boiler tubes and melts, it can help initiate hot corrosion.

This problem of coal-based power plant in boilers due to burning of coal is known as fireside corrosion. The other corrosion related problems in such power plants are from the inside of a boiler tube, which is called a steamside corrosion or waterside corrosion problems, which are due mainly to poor control of water chemistry.

In addition, there are other possible areas in coal-based power plants which are prone to corrosion. The following are specific areas typical of power generating facilities that are most vulnerable and in need of protection: (a) Ducts such as flue gas inlet duct, scrubber outlet ducts, and bypass ducts; (b) stacks, steel liners, and brick liners; (c) fuel handling areas; (d) scrubber module (quencher, absorber); (e) demineralized water areas, collection sumps/neutralization basis.

Before discussing in detail the three main corrosion problems, fireside, steamside and condenser related problems, it is first

Table 2:   Typical values of S, moisture content, C, bulk density and ash content in some Indian coals.[11]

Typical Sulfur Content in Coal	Anthracite Coal : *0.6–0.77 weight %* Bituminous Coal : *0.7–4.0 weight %* Lignite Coal : *0.4 weight %*
Typical Moisture Content in Coal	Anthracite Coal : *2.8–16.3 weight %* Bituminous Coal : *2.2–15.9 weight %* Lignite Coal : *39 weight %*
Typical Fixed Carbon Content in Coal	Anthracite Coal : *80.5–85.7 weight %* Bituminous Coal : *44.9–78.2 weight %* Lignite Coal : *31.4 weight %*
Typical Bulk Density of Coal	Anthracite Coal : *50–58 (lb/ft^3), 800–929 (kg/m^3)* Bituminous Coal : *42–57 (lb/ft^3), 673–913 (kg/m^3)* Lignite Coal : *40–54 (lb/ft^3), 641–865 (kg/m^3)*
Typical Ash Content in Coal	Anthracite Coal : *9.7–20.2 weight %* Bituminous Coal : *3.3–11.7 weight %* Lignite Coal : *4.2 weight %*

important to see the schematic of a coal-based power plant, which starts from the heap of coal blocks, which are then pulverized and converted to fine powder. These are then injected through burners where it is burned in air. Heat produced is taken by boilers which convert an input of well cleaned water into steam. The steam so produced is expanded through a turbine, which turns a generator. The steam at the low pressure exit end of the turbine is condensed and returned to the boiler. This is shown schematically in Fig. 3.

4.1 *Fireside Corrosion*

As illustrated in Fig. 4, in fossil fuel fired steam power plant, there are three fluid flow loops circulating through the system: fuel–air, water–steam, and condenser cooling. In the fuel–air loop, the fossil fuel is burnt in air and transfers its heat to a series of heat exchangers. In the water–steam loop, clean feed water is converted into superheated steam in a boiler, which expands through a series of turbines, converting its heat into mechanical energy. In the condenser cooling loop, cold water is passed through the condenser and can be

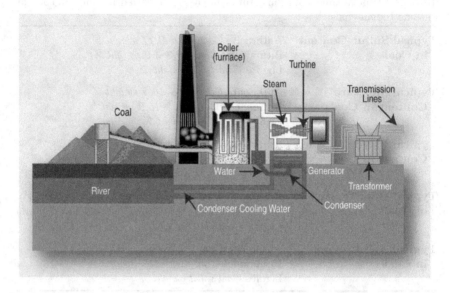

Fig. 3.   Systematic working of a coal-based power plant.[12]

re-circulated or is exhausted back to the source of the cooling water.
Each fluid loop possesses its unique corrosion problems.

The fossil fuel is burnt in a very large chamber constructed of
water walls (consisting of vertical or spiral tubes welded together
in a web), where the feed water is heated. In subcritical boilers,
the saturated steam is superheated in tubular heat exchangers. In
supercritical boilers, the liquid becomes superheated vapor without
undergoing a phase change.

When coal particles are introduced into the flame, the moisture
and the volatile species are driven off; the fixed carbon in the
individual particles begins to burn. The contained mineral matter
may be melted or vaporized, and is largely oxidized. The sulfur-
containing compounds in the coal (such as FeS) are converted to
oxides such as Fe_2O_3, K_2O, Na_2O, SO_2, and SO_3. The relative
proportions of SO_2 and SO_3 in the flame depend on the available
oxygen and the temperature. SO_2 is thermodynamically favored
at higher temperatures ($>70°C$); the formation of SO_3 can be
catalyzed by certain metal oxides, so that the proportion of SO_3

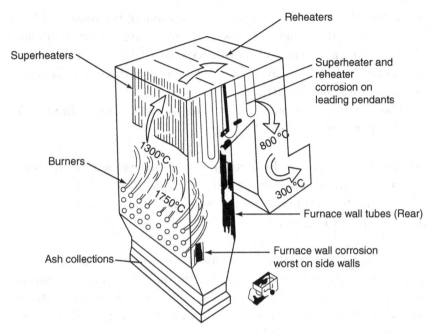

Fig. 4. Schematic of coal-fired boiler.[13]

Table 3: Showing the extent of corrosion as a function of total Na and K content in water.[14]

Water Soluble Na + K (wt.%)	Corrosiveness
<0.5	Low
0.5–1.0	Medium
>1.0	High

in flue gas may increase downstream of the burners. Thus, the gaseous species released, as the coal passes through the flame, contain potential corrodent such as sulfur, vapor of alkali metal salts, and chlorine compounds (mostly HCl). The quality of coal used is very important. Raask[14] has proposed a simple three category ranking of the corrosiveness of coals based on the sum of the percentages of water-soluble sodium and potassium in the coal:

A correlation has been found between the corrosion rate of super heater/re-heater tubes and the chloride content of the coal.[3] Chloride (more than 0.2 wt%) has been found to promote the release of both Na and K into the flame and acts as a strong catalyst for the molten trisulfate attack. There is also evidence that HCl formed in the flame can destroy the Fe_2O_3 layer on a steel surface, thereby exposing it to additional oxidative attack.[15]

The fireside corrosion of various components of a coal-fired boiler may be attributed to the following:

(a) Reducing (sub-stoichiometric) conditions caused by impingement of incompletely combusted coal particles and flames,
(b) Accelerated oxidation from overheating, and
(c) Molten salt or slag-related attack.

The fireside corrosion is generally localized to regions on the walls near the burners. The thick, hard, external scales formed often exhibit cracks which resemble an alligator hide. Reducing atmospheric corrosion can result due to direct reaction of the water wall tubes with a sub-stoichiometric gaseous environment containing sulfur, or with partially combusted char containing FeS.[16–19] The reducing conditions have two main effects on corrosion. First, they tend to lower the melting point of any deposited slag, increasing its ability to dissolve the normal oxide scales, and second, the stable gaseous sulfur compounds under these conditions include H_2S, which is more corrosive than SO_2 that predominates under oxidizing conditions. Analysis of cross-section of a 1.25 Cr-alloy waterwall tube removed from a boiler where it was exposed to "reducing" atmosphere, confirmed the scale, a mixture of sulfide and oxide, suggesting that the conditions in the combustion gas at the waterwall were close to those suggested by the Fe_3O_4/FeS boundary in the phase stability diagrams.[20] Overheating of the super heater is related to the poor design of the boiler, when slagging problems are experienced.

Change in parameters such as the feed rate of coal to attain the desired steam temperature can cause overheating. Overheating of the reheaters can trigger rapid start up situations, when the combustion gas temperature at the reheater reaches its maximum value before full steam flow through the reheater is achieved. The overheating leads to accelerated oxidation of both the fireside and the steamside surfaces of the tubes to produce thickened, hard scales. Above 570°C, a very non protective scale of wustite (FeO) can be formed on iron which leads to the onset of rapid oxidation.

4.1.1 Mechanism of molten salt or slag-related attack

Molten salt or slag-related attack takes several forms. Local disruption of the normal oxide film on the wall tubes can lead to either accelerated oxidation, or to oxidation/sulfidation attack due to sulfur species in the slag. Alkali sulfates, deposited on the waterwalls, may react with SO_2 or SO_3 to form pyrosulfates such as $K_2S_2O_7$ and $Na_2S_2O_7$, or possibly complex alkali-iron trisulfates, the latter compounds being formed in thicker deposits after long times at about 482°C. The K_2SO_4–$K_2S_2O_7$ system forms a molten salt mixture at 407°C when the SO_3 concentration is 150 ppm. The above mechanism may be depicted by the following sequence of reactions:

$$K_2SO_4(s) + SO_2(g)/SO_3(g) = K_2S_2O_7(s) \quad (407°C, 150 \text{ ppm } SO_3).$$

This result in a molten salt mixture ($K_2SO_4 + K_2S_2O_7$) can directly react with the heat exchanger tubes and regenerate potassium sulfate:

$$K_2S_2O_7(s) + 3Fe(s) = Fe_2O_3(s) + K_2SO_4(s).$$

By such a mechanism, the pyrosulfate can react aggressively with any protective iron oxide scales on the tubes, and lead to accelerated wastage through fluxing of the oxides and attack of the substrate metal. The corresponding sodium system can become liquid at 400°C with about 2500 ppm of SO_3. Such high concentrations of SO_3 is possible in the stagnant regions, beneath deposits. Thus, a similar attack by $Na_2S_2O_7$ may occur when a high-sulfur coal produces combustion gases containing high levels of sulfur oxides. However, as it is well known that the levels of SO_3 present at this location in a boiler, burning a typical coal are such that $K_2S_2O_7$ is unlikely to be found at temperatures above about $510°C$, and $Na_2S_2O_7$ only up to about $400°C$.

Deposit-related molten salt attack of the pendant tubes concerns the development of conditions beneath a surface deposit which are conducive to the formation of a low melting salt of the type $(Na, K)_3Fe(SO_4)_3$. Catalytic oxidation of SO_2 in stagnant zones beneath a layer of deposit can lead to nearly equilibrium levels of SO_3, so that conditions are favorable for the formation of trisulfates in deposits up to about $704°C$. Above this temperature, the required SO_3 concentrations cannot be sustained, and the trisulfates become unstable, decomposing to the alkali sulfates which are solid. There is wide acceptance that compounds of this type play a critical role in the corrosion of superheater tubes. Deposits on the superheater tubes are usually found to be tightly bonded to the tubes at the room temperature. They typically consist of three distinct layers:

(a) A hard, brittle and porous outer layer, which is the bulk of the deposit and has a composition similar to the boiler fly ash.
(b) A white intermediate layer. When this layer has a chalky consistency, corrosion is found to be mild or non-existent. When it is fused and semi-glossy, corrosion is found to be severe. Compounds identified in this layer include complex alkali sulfates and the alkali-iron trisulfates.
(c) A glossy inner layer composed primarily of oxides and sulfides of iron. The typical appearance of a corroded superheater tube is illustrated in Fig. 5. Tile thickened, non-protective scale formed beneath such deposits comprises of mixed layers of iron oxides and sulfides as is evident from Fig. 6.

5. Fireside Erosion Problems

Light erosion damage is usually manifested by polishing of the affected surface. The eroded area is often quite clean and free of deposits. Thinned and flattened areas result from more severe erosion. Erosion tends to be localized to particular areas of the boiler,

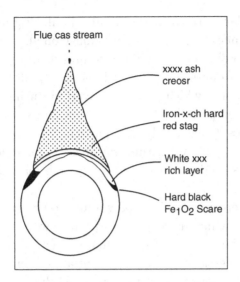

Fig. 5.

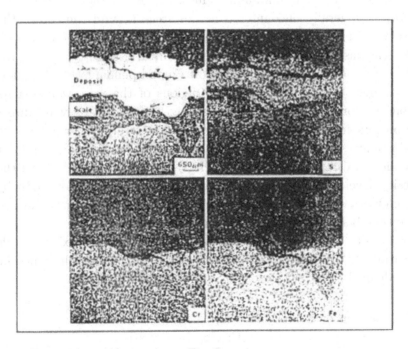

Fig. 6.

and to particular parts of a given tube bank. Figure 8 shows the typical appearance of fly ash erosion damage on all economizer tubes. The factors in coal which contribute to fireside erosion problems are large particles of dense minerals such as quartz, or FeS_2, and those mineral constituents which may be converted during combustion to hard/abrasive compounds such as alumina and silica based oxides. System variables (such as fly ash particle velocity and the angle of impingement), and operating variables (which define the size, shape, hardness and density of the fly ash particles), are very important.

The rate of erosion loss is usually found to be proportional to the impact velocity, which usually varies from power of velocity of 2 to 4, and also on the number of individual impacts.

Erosion damage is, therefore, a potential problem at any point on the fireside of the boiler where the ash laden flue gas contacts boiler tubes or internal support structures at velocities and with particle loadings above some minimum values. Fly ash erosion of waterwall tubes is generally encountered in the areas around the top of the rear wall of the furnace, where the flue gas is turned to flow through the rear pass. Erosion results largely from turbulence created by the change in flow direction, and by flow around pendant tube bundles.

Other types of erosion in the regions of the waterwalls occurs from ash or slag entrained by wall blowers, or possibly by direct impingement of the flames. The erosion is rarely the cause of the tube failures in outlet superheater/reheater tubes. Erosion can occur in these areas when the gas velocity is locally increased above the design level (15–20 m/s). Blockage of the normal gas flow path by slag or ash deposits can result in channeling of the flue gas through the tube banks, with rapid thinning of the tubes.

Raask[13] developed the concept of an "Erosive Index" for coals relating the erosivity to the quartz content of the fly ash, which he has defined as

Erosion by fly ash index (I)

$$= \frac{\text{Erosion by Fly ash}}{\text{Erosion by equal weight of } 100\,\mu\text{m quartz particles}}.$$

The value of I was found to range from 0.2 to 0.4. From experimental results, erosion rate of mild steel by $100\,\mu m$ quartz grains is given as

$$W_e = 9.5 \times 10^{-10} \times W_m \times U^{2.5},$$

where W is weight of the metal eroded, W_m is weight of impacting particles (Kg), U is velocity of the particles (m/sec).

The corresponding expression for erosion by fly ash is

$$W_e = 9.5 \times 10^{-10} \times I \times W_m \times U^{2.5}.$$

Using these calculations, one can say that at velocities around $35\,m/s$, it may cause erosion by fly ash which can cause the tube failure in 10,000–50,000 h, which is in good agreement with practical experiences (usual design velocities are 15–20 m/s). Kratina[20] has proposed an erosion prediction method based on "Coal Erosiveness Factor" (CEF) which is determined as follows:

$$CEF = 8.25/HHV \times (\% \text{ ash}) \times a \, (\% \text{ erodents}),$$

where 8.25 is a constant related to unit heat input, HHV is fuel heat value (in B ThU) and a, is the erosion index of the erodents present. The erosion index is determined as follows:

(% erodents)

$$= a1 \, (\%\text{quartz}) + a2 \, (\%SO_2) + a3 \, (\% \, Al_2O_3) + a4 \, (\% \, Fe_2O_3).$$

Interpretation of CEF value for tube bank velocities in the range of 16–21 m/s are as follows:

CEF predicted erosion value:

0–0.5 — No erosion problem;
0.5–1.0 — Mild to persistent erosion;
1.0–1.5 — Serious to very serious erosion;
1.5– — More severe erosion.

The complete elimination of oxidation and hot corrosion/erosion problems encountered in coal-fired boilers is not possible considering the complexity of the environments and the mechanisms involved. However, with proper selection of construction materials and design of components, modification of operating conditions, use of good quality coals, and applications of various heat-resistant coatings, the problems can be effectively checked to a great extent.

6. Steamside Corrosion Problems

Steamside corrosion is basically due to the corrosion of boiler tube from inside due to bad water quality. In a one liner, the steamside corrosion can be controlled using a water of controlled chemistry. Various factors responsible for corrosion on steamside are

(a) Presence of oxygen in water.
(b) Presence of chlorides, a natural impurity in water.
(c) Total dissolved solids, which precipitate at higher temperatures.
(d) pH.

Effect of these parameters varies based upon different circuits in the steamside which can be described as follows:

(a) Feed water circuits, pumps and valves;
(b) Boiler internal surface;
(c) Condensate Circuit;
(d) Turbine.

The most critical is the feed water introduced into the boiler. The feed water basically is a combination of 100% make up water or a combination of those retained from condenser and makeup water. The main problems are to reduce oxygen level, chlorides, dissolved solids and adjust the hardness.

Normal river and ponds have about 18% of dissolved oxygen at normal temperatures. This oxygen is harmless at ambient temperatures to internal walls of boiler tubes, but can cause severe corrosion attack at high temperatures when present in boiler water or steam. Figure 7 shows an oxygen attack on a boiler tube.

Fig. 7.   Oxygen attack of a boiler tube — from *Corrosion Atlas.*

Thus oxygen must be completely removed from the feed water. It has been found that the concentration of oxygen in feed water must be lower than 5 parts per billion (ppb). This can be achieved by a series method such as inert gas pumping, however, the last trace of oxygen is removed by addition of scavengers such as sodium sulfite (Na_2SO_3) or hydrazine (N_2H_4):

$$Na_2SO_3 + O_2(g) = Na_2SO_4.$$

About eight parts by weight of Na_2SO_3 is used and the reaction is catalyzed by cobalt salts at pH between 5 and 8 at elevated temperatures. The Na_2SO_3 should contain about 0.25% of cobalt sulfate. The other scavenger is hydrazine which removes oxygen by the following reaction:

$$N_2H_4 + O_2(g) = N_2 + H_2O.$$

One part of hydrazine drives out one part of oxygen through nitrogen gas. This direct reaction proceeds rather slowly at around 150°C, while at temperatures as low as 70°C, the reaction proceeds rapidly by following indirect reaction:

$$N_2H_4 + 6Fe_2O_3 = 4Fe_3O_4 + N_2 + H_2O,$$

$$4Fe_3O_4 + O_2(g) = 6Fe_2O_3.$$

Such reaction proceeds on ferrous metal. Steel tubes are the right places for such reaction. The only limitation of hydrazine is that it is a carcinogenic material and therefore has limited utility as boiler water scavenging material.

The other treatment which is essential for feed water is the prevention of deposits (scales, sludges, and corrosion products) on the waterwalls, which result in poor heat transfer. The thicker the deposit, the higher the temperature required for heat transfer. Three types of deposits which may form on waterwalls are as follows:

1. Carbonic acid corrosion

$$Fe + 2H_2CO_3 = Fe^{2+} + H_2 + 2HCO_3^-.$$

2. Reaction with oxygen

$$4Fe + 6H_2O + 3O_2 = 4Fe(OH)_3.$$

3. Corrosion product with bicarbonate

$$2Fe(HCO_3)_2 + \frac{1}{2}O_2 = Fe_2O_3 + 4CO_2 + 2H_2O.$$

In order to fix these deposits, following treatments are recommended:

(a) Phosphate Treatment to control pH and dissolve scale

Ca^{++} and Mg^{++} ions are mostly suspended in the water as dissolved salts and can precipitate and deposit on waterwalls. These need to be removed. Phosphate treatment which either consists of treating water with phosphoric acid and NaOH or using a treatment with sodium tri-phosphate is carried out. Ca^{++} and Mg^{++} ions get converted to respective phosphates and precipitated in sludge form and thus easily removed by blow down:

$$10Ca^{++} + 6PO_4^{--} + 2OH^- = 3Ca_3(PO_4)_2 \cdot Ca(OH)_2,$$
$$\text{(Calcium hydroxyapatite)}$$
$$2Mg^{++} + 2OH^{--} + 2SiO_2 + H_2O = MgSiO_2 \cdot Mg(OH)_2 + H_2O,$$
$$\text{(Serpentine)}$$

pH in this method is maintained between 10 and 11.2.

In case of addition of trisodium phosphate, it hydrolyzes to produce hydroxide ions and phosphate acts as buffer to minimize caustic cracking.

$$Na_3PO_4 + H_2O = Na_2HPO_4 + NaOH.$$

No free caustic is maintained in the boiler water. A ratio of 3 Na to 1 PO_4 is maintained by phosphate addition.

Another method of controlling the deposits is by addition of chelants which form soluble complexes with metal cations. Some common chelants are ethylenediaminetetraacetic acid (EDTA) and nitro triacetic acid (NTA):

$$Ca^{++} + EDTA = CaNa_2EDTA + NaOH.$$

By this formation of calcium carbonate is eliminated, provided the total feed water hardness be lower than 2.0 ppm, low ferric ions and oxygen be very low.

7. Waterside Corrosion Problems

The waterside corrosion problems in boilers can be controlled mainly by forming a good magnetite coating on the normally used steel walls of the boiler and its maintenance. The applied coating should be least permeable to water because water is the species responsible for the continuous damage of the thin magnetite film formed on the steel during the operation. Purity of feed water is also an important factor from the corrosion point of view. Hence, the chemistry of feed water should be monitored and efforts should be made to reduce the quantity of dissolved gases and salts in it.

Thus, the main corrosion problems in boiler water are oxygen attack, caustic corrosion, hydrogen damage, stress corrosion cracking and corrosion fatigue.

The boiler fluid may become acidic or caustic, depending on the presence of corrosion deposits and flow interruptions. Under acidic conditions, the steel boiler tubes may be hydrogen embrittled; under caustic conditions, the tubes may be caustic gouged. The corrosion in a steam boiler may be represented by the equation,

$$3Fe(s) + 4H_2O(l \text{ or } g) = Fe_3O_4 + 4H_2(g).$$

From the corrosion point of view, a boiler is nothing but a thin film of magnetic iron oxide supported by steel. This oxide film is continuously damaged and repaired during boiler operation, with simultaneous production of hydrogen. The superheater and reheater tubes suffer from steam oxidation of the inner surfaces and hot corrosion of the outer surfaces. The fireside corrosion is a typical problem. In coal-fired boilers, it exhibits a maximum rate at 700–750°C, where the corrodent is a liquid, and decreases to a minimum at higher temperatures.

8. Corrosion Control Methodologies for Power Plants

Fireside corrosion is mainly due to the presence of impurities in coal. Thus, reducing these impurities can be one of the ways, for example, blending of coal types and washing of coal can improve the stoichiometric balance of the coal and air flow to each burner. Standard coal washing can remove approximately one-half of the sulfur and alkali metal content of the coal. Blending a known corrosive coal with another also helps to produce a less corrosive ash. Another way is to play with operating parameters such as by limiting the maximum temperature of steam generated to about 540°C. For corrosion protection of superheater/reheater, the tube metal temperature should be maintained in a regime where the rate of corrosion from alkali-iron sulfate-type attack is considerably less than the maximum possible. However, under normal circumstances, there is little freedom to change the operating conditions. However, sometimes the changes are not very effective. Thus, the above modifications in operational parameters do not provide a long-term solution to corrosion.

9. Corrosion Protection Methods

9.1 *Selection of Materials*

Most of the power plant is usually made of steel, pressure vessel, boiler tubes, and heat exchanger tubes turbine blades. Since normal steel starts deteriorating both in mechanical properties as well as corrosion properties, it is modified to meet these requirements. The main components where there is possibility of corrosion is boiler tubes, as discussed above, these need protection both from steamside as well as from fireside. As discussed above, the waterside problem is usually solved by controlling the chemistry of the water, such as oxygen level, chlorides, pH and hardness. Further boiler tubes are divided into superheaters and economizers and reheater. The main difference here is in the temperature while superheaters who carry steam at highest temperature need better alloy. The most important choice of boiler materials is 21/4Cr–1Mo steel for reheaters and economizers and 9Cr-1Mo for superheaters. Cr is for corrosion resistance while Mo helps to take care of strength as well as creep. Cr helps in making a Fe–Cr spinel oxide for 21/4Cr–1Mo steel, while a Cr-rich spinel is formed for 9Cr–1Mo steel which protects the reheater and superheater tubes from corrosion from waterwall side.[21,22]

From the fireside corrosion, both 21/4Cr–1Mo and 9Cr–1Mo steel corrode badly if the coal is having sulfur and salts. Both oxidation & sulfidation can take place. And in case, ash deposit takes place on superheater tubes, there is possibility of hot corrosion. Both sulfidation and hot corrosion problems can be minimized or overcome by the application of Cr-rich coatings on the outside of superheater tubes (fireside). These coatings are either stainless steels or Ni–25Cr coating, applied by thermal spray. Such a coating protects the heat exchanger tubes from any high temperature corrosion.

The other way to apply such coatings is by laser surface modification.[23-30] Ni-25Cr alloy is applied on the superheater tubes by laser cladding. Figure 8(a) shows a laser cladded heat exchanger tube applied by a Nd:YAG laser. The coated heat exchanger tubes were then kept in the boiler along with other tubes which were uncoated. The tubes were removed after one year of exposure. The uncladded tubes showed severe corrosion extending into the thickness of the coating, while cladded tubes showed no damage at all. This is shown clearly in Fig. 8(b).

Another important part of power plant is the material selection for steam turbine, which basically means on the design and whether it is sub-critical, supercritical or ultra-critical.[31-34] The requirement is based upon temperature and pressure. The choice of casing material, valves, bolts, rotor discs very much depends on that. The choice of blading material will depend on (i) the temperature of the rotor, hence on the thermal expansion characteristics of the material, from which it is made, and (ii) the size and shape of the blade, which will be designed using computational fluid dynamics modeling. There will be a requirement for the generation of data on the interaction of these materials with steam; results from recent research suggest that it will be important to have higher-Cr levels in these alloys to avoid preferential internal attack in steam.[35-37] Tables 4 and 5 below gives a list of various materials used for steam turbine application including pipings.

While it is not known if solid particle erosion from entrained particles of oxide scale that may be exfoliated from the superheater and reheater tubing will be a greater problem than encountered in current steam turbines, it will be prudent to ensure the availability of erosion-mitigating coatings that are compatible with the high temperature blading materials. However, Mg-based additives, such as MgO and Mg $(OH)_2$, have been tried in some cases to prevent corrosion by sulfuric acid condensation in the cold end of the boiler. Also, Rahmel[40] reported that addition of either Mg or Ca-sulfate reduced the corrosion of stainless steels caused by K_2SO_4. CaO has also been used in some cases.

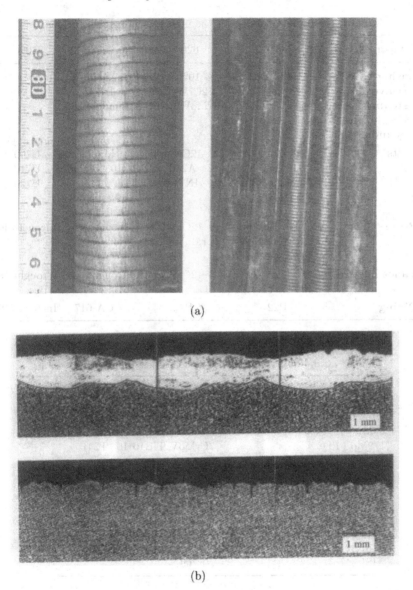

Fig. 8. (a) Laser cladded heat exchanger tubes with Ni–25Cr alloy. (b) Uncladded tube showing deep cuts in the cross-section of a heat exchanger tube.

Table 4: Material selection for the high pressure steam turbine.[38]

Component	566°C	620°C	700°C	760°C
Casing/Shells (valves; steam chests, nozzle box; cylinders)	CrMo (cast 10CrMoVNb)	9%–10%Cr(W) 12CrW(Co) CrMoWVNbN	Inconel 625 IN718 Nimonic 263	Inconel 740
Bolts	422 9-12CrMoV Nimonic 80A IN718	9–12CrMoV A286 IN718	Nimonic 105Nimonic 115 Waspaloy IN718	U700 U710 U720
Rotors/Disc	1CrMoV 12CrMoVNbN 26CNiCrMoV115	9–12CrWCo 12CrMoWVNbN	Inconel 625 Haynes 230 Inconel 740	Inconel 740
Vanes Blades	422 10CMoVNbN	9–12CrWCo	Wrought Ni Base	Wrought Ni Base
Piping	P22	P92	CCA 617	Inconel 740

Table 5: Candidate materials for boiler tubes, pipes and headers.[39]

Materials for Boiler tubes and pipes	
Sub-critical	Supercritical
C-Mn	HCM25(T23)
1/2Mo (T1)	7CrMoV TiB10 10 (T24)
11/4Cr1/2MoSi (T11)	X20 CrMoV 12 1
21/4Cr1Mo (T22)	X10CrMoVNb 91 (T91)
X20 CrMoV 12 1	X10CrMoWVNb 911 (E911)
X10CrMoVNb 91(T91)	X10CrMoWVNb 92 (T 92-NF616)
AISI 304	X10CrMoVNb 12 1 (T122)
AISI 316	X8CrNNiMoVNb 16 13
AISI 321	X3CrNiMoNb 16 16
AISI 347	NF709

10. Summary

Corrosion problems in coal-based power plants are basically due to impurities present in the coal. Presence of sulfur, salts and silica cause erosion and corrosion of superheater tubes, leading to hot corrosion, sulfidation and erosion. Waterside or steamside corrosion problem arise only due to wrong water chemistry control. Once the water is properly treated by removing oxygen, chlorides and maintain proper dissolved solids, steamside problems are minimum. Most of the materials used for economisers and superheated tubes are 214Cr–1Mo and 9Cr–1Mo steels, respectively. These materials though are perfectly O.K. for steam side and waterside corrosion, but additional chromium rich coatings or claddings are needed to protect fire side corrosion. For supercritical power plants, the low alloy steels especially 9Cr-1Mo steels are further modified by alloying with W and Nb to take care of higher temperature of 620°C and higher pressure of 30 bar.

References

1. Thermal Power Station. *Wikipedia*, Wikimedia Foundation. Retrieved from http://en.wikipedia.org/wiki/Thermal_power_station.
2. *Hydropower*, Salem: Legislative Committee Services, 2011.
3. Gujarat Cleaner Production Centre — Envis Centre Report. *CLEANER PRODUCTION OPPURTUNITIES*. Retrieved from *http:// gcpcenvis.nic.in/Experts/Therma Power Plants. Pdf.*
4. "Coal". Retrieved from http://en.wikipedia.org/wiki/Coal.
5. G.M. Johnson *et al.*, Distribution of sulfur species in the burnt gas of fuel-rich propane-air flames. *Combustion and Flame* **15**(2) (1970), pp. 211–214.
6. A.S. Khanna, *Introduction to High Temperature Oxidation and Corrosion* (ASM International, Ohio, 2002).
7. A.S. Khanna and S.K. Jha, Degradation of materials under hot corrosion conditions, *Trans. Indian Inst. of Metals* **51**(5) (1998), p. 279
8. W.J. Quadakkers *et al.*, Investigation of the corrosion mechanisms of nickel and nickel-based alloys in SO_2-containing environments using an evolved gas analysis technique, *Mater. Sci. Eng. A* **120**(1989), pp. 117–122.
9. W.J. Quadakkers *et al.*, Influence of sulphur impurity on oxidation behaviour of Ni–10Cr–9Al in air at 1000°C, *Mater. Sci. Tech.* **4**(12) (1988), pp. 1119–1125.
10. A.S. Khanna *et al.*, The influence of sulphur and its interaction with yttrium on the composition, growth and adherence of oxide scales on alumina forming alloys, in *The Role of Active Elements in the*

Oxidation Behaviour of High Temperature Metals and Alloys (Springer Netherlands, 1989), pp. 287–297.

11. *Classification of Coal.* Retrieved from http://www.engineeringtoolbox. com/classification-coal-d_164.html.

12. Fossil Fuel Power Station, *Wikipedia,* Wikimedia Foundation.

13. A.S. Khanna, Fireside corrosion and erosion problems in coal based power plants, in *National Workshop on Boiler Corrosion,* NML, Jamshedpur, 1995.

14. E. Raask, *Mineral Impurities in Coal Combustion* (Hemisphere Publishing Corp., Washington, D.C., 1985), pp. 342–343.

15. D.J. Lees, Selection of corrosion resistant coatings for boiler tube applications, *Surface Treatments for protection,* **10**, 1978, pp. 174–182.

16. P. Kratina and J. McMillan, Fly ash erosion in utility boilers-prediction and protection, *Canadian Electrical Association Conf.,* NACE International Conference, Chicago, 1980.

17. W.T. Reid, External corrosion and deposits: Boilers and gas turbines (1971).

18. R.C. Corey, B.J. Cross and W.T. Reid. External corrosion of furnace wall tubes. II. Significance of sulfate deposits and sulfur trioxide in corrosion mechanism, *Trans. ASME.* **67**, (United States, 1945).

19. S. Phillips *et al.,* Application of high steam temperature countermeasures in high sulfur coal-fired boilers. *Retrieved from http:// www.hitachi.powersystems.us/supportingdocs/forbus/hpsa/technical_ papers/EP2003B.pdf* (2003).

20. I.P. Ivanova, V.P. Kaminski and A.G. Belyaeva, High-temperature corrosion of waterwall tubes in supercritical boilers burning anthracite fines, *Ther. Eng+.* **19**(1) (1972), pp. 15–18.

21. A.S. Khanna, J.B. Gnanamoorthy and P. Rodriguez, Oxidation behavior of 21/4 Cr-1Mo and 21/4 Cr-1Mo-Nb steels, *T. Indian I. Metals* **40**(6) (1987), pp. 503–513.

22. A.S. Khanna, P. Rodriguez and J.B. Gnanamoorthy, Oxidation kinetics, breakaway oxidation, and inversion phenomenon in 9Cr-1Mo steels, *Oxid. Met.* **26** (3,4) (1986), pp. 171–200.

23. M. Li *et al.,* High-temperature oxidation resistance improvement of titanium using laser surface alloying, *J. Mater. Sci.* **30**(20) (1995), pp. 5093–5098.

24. M.G. Pujar, *et al.,* Effect of laser surface melting on the corrosion resistance of chromium-plated 9Cr-1Mo ferritic steel in an acidic medium, *J. Mater. Sci.* **28**(11) (1993), pp. 3089–3096.

25. K. Sridhar *et al.,* Formation of highly corrosion resistant stainless steel surface alloys for marine environments by laser surface alloying, NACE International Conference, Texas, 1998.

26. A.S. Khanna and K. Sridhar, Corrosion and oxidation behavior of laser treated surfaces, ASM International, Member/Customer Service Center, Materials Park, OH 44073-0002, USA, 1998, pp. 395–429.

27. K. Sridhar *et al.*, Laser surface alloying of type 304 stainless steel for enhanced corrosion resistance (1996).

28. A.S. Khanna, S. Mahapatra and A. Gasser, Characterization oxidation and sulphidation resistance of UNS7718 superalloy fabricated by laser forming process, *Corrosion 2002* (2002).

29. A.S. Khanna, R. Streiff and K. Wissenbach, Formation of high chromium surface alloys on 2.25 Cr–1Mo and 9Cr–1Mo steels using a single step laser treatment to improve their high temperature oxidation corrosion resistance (1999).

30. S. Kumari, A.S. Khanna and A. Gasser, The influence of laser glazing on morphology, composition and micro hardness of thermal sprayed Ni-WC coatings (2006).

31. P. Mathiazhagan and A.S. Khanna High temperature oxidation behavior of materials for supercritical fossil fuel power plant in air and O2 + water vapor environments.

32. P. Mathiazhagan and A.S. Khanna, High temperature oxidation behavior of materials for supercritical fossil fuel power plant in air and O.

33. P. Mathiazhagan and A.S. Khanna, Oxidation behavior of power plant materials at different water vapour environments.

34. P. Mathiazhagan and A.S. Khanna, Effect of water vapor on the oxidation behavior of modified low alloy steels at high temperatures, *Arab. J. Sci. Eng.* **34**(2) (2009), p. 159.

35. A.S. Khanna and P. Kofstad, Effects of water vapor on oxide growth on 304 stainless steel at 900°C, *Microscopy of Oxidation* (1990), pp. 113–118.

36. A.S. Khanna and P. Kofstad, Effect of temperature pressure and flow rate on the oxidation of 304 stainless steel in dry and wet oxygen, *Met. Mater. Processes* **1**(3) (1989), pp. 177–195.

37. K.M.N. Prasanna, A.S. Khanna and W.J. Quadakkers, Effect of water vapor on the oxidation of FeCrAl-and NiCrAl-base ODS alloys (1998).

38. I.G. Wright *et al.*, Materials issues for turbines for operation in ultra-supercritical steam, research sponsored by the US Department of Energy, Office of Fossil Energy, Advanced Research Materials Program, under Contract DE-AC05-00OR22725 with UT-Battelle, LLC (2004).

39. K. Singh, *BHEL J.* **27**(2) (2006), pp. 1–19.

40. A. Rahmel, Influence of calcium and magnesium sulfates on the high temperature oxidation of austenitic chrome-nickel steels in the presence of alkali sulfate and sulfur trioxides, *Proc. Int. Conference*, 1963.

Chapter 6

High Temperature Corrosion Problems in Aircrafts

A.S. Khanna* and Vinod S. Agarwala[†]

*Department of Metallurgical Engineering and Materials Science
Indian Institute of Technology Bombay
Mumbai 400076, India

[†]Senior Staff Scientist and US Navy Esteemed Fellow (Ret.),
Technical Consultant, Iron Pillar Engineers
1600 Green Street, Philadelphia, PA 19130, USA

1. Introduction

Discovery of jet engines necessitated the need for stronger materials which can sustain high strength at higher temperatures. This innovation coincided with the discovery of the existence of tiny coherent particles in an austenitic matrix by Bradley and Taylor in 1940 though the enhanced creep resistance achieved by adding small amount of Ti and Al in a 80Ni–20Cr alloy was already proved by Bedworth and Pilling in 1929 and also simultaneous research in England, United States, and Germany, in 1930s, succeeded in creating strong alloys of nickel and iron base, containing Cr and Y, plus carbide-strengthened cobalt alloys. These tiny particles which strengthened Ni-base matrix had same structure as the base matrix and therefore designated as austenite prime (γ'). Further, major jump in aerospace material development took place in 1950, when vacuum melting process was discovered. It removed undesirable alloy impurities that affected the alloys developed in the '30s and '40s. It also permitted additional alloying elements and control of vital reactive strengthening and oxidation-resistant elements. The total

129

alloy chemistry was improved immensely, and complex cast shapes were possible. Before that, the superalloys were made using air melting which used to bring lots of ceramic inclusions such as oxides due to oxidation of elements such as aluminum and chromium.

Aircraft corrosion can be classified into low temperature and high temperature corrosion. The development of materials for aircraft airframe has been the most demanding. It coincides with one of the first systematic efforts on the development of the high-strength aluminum alloys, which were considered as the first choice for airframes and other structural parts of a transport aircraft, mainly due to their low density and high corrosion resistance. Alloys primarily utilized are 2024-T4, higher strength alloys, viz., (2014-T6, 7075-T6, 7079-T6, and 7178-T6). Where sheet is used, the Al-clad form is preferred. The upper skins and spar caps of wings often are of 7075-T6 and 7178-T6, with a critical requirement of high compressive strength, rather than critical in tension or fatigue. For wing tension members, shear webs, and ribs, alloys 2014-T6, 2024-T4, and 7075-T6 are used extensively, where fatigue performance and fracture toughness, combined with high strength, are the main characteristics. All these materials experience general corrosion problems classified into low temperature corrosion. Readers are requested to refer to an extensive review in Ref. 1.

This chapter is mostly devoted to high temperature corrosion and the most important component of an airplane, prone to high temperature corrosion, is gas turbine. A typical gas turbine sketch is shown in Fig. 1. It consists of three main sections: compressor, combustion chamber and turbine.

The working fluid (mainly aviation fuel) is initially compressed in the compressor. It is then heated in the combustion chamber. Finally, it goes through the turbine, which converts the energy of the gas into mechanical work. Part of this work is used to drive the compressor and the remaining part is known as the network of the gas turbine. The main functions of a gas turbine are schematically illustrated in Fig. 2.

The running of a gas turbine depends on Joule–Brayton cycle, which consists of four important points. It starts at position 1, where

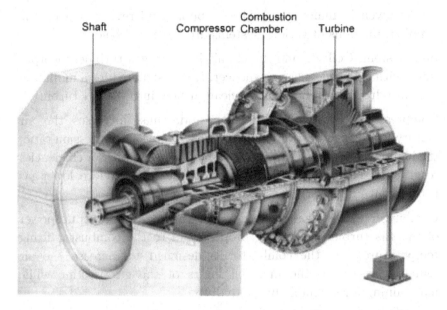

Fig. 1. A typical gas turbine sketch.

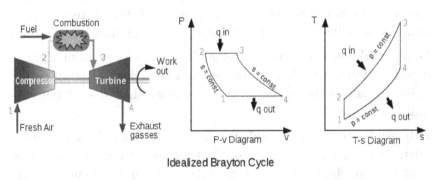

Idealized Brayton Cycle

Fig. 2. Main functions of a gas turbine shown along with Brayton Cycle.[2]

the gas has passed through the inlet, after that the gas passes through the compressor. It is assumed that the compression is performed isentropically, as shown from point1 to point 2. The gas is then heated in the combustor (Point 3). This is done isobarically (at constant pressure). So, p2 = p3. Finally, the gas is expanded in the turbine (Point 4). This is again done isentropically. So, s3 = s4.

As given in many standard text books and review papers, the detailed functions of various systems are described below:

Inlet system: Collects and directs air into the gas turbine, incorporated with an air cleaner and silencer. It is designated for a minimum pressure drop while maximizing clean airflow into the gas turbine.

Compressor: Provides compression, and, thus, increases the air density for the combustion process. The higher the compression ratio, the higher the total gas turbine efficiency. Low compressor efficiencies result in high compressor discharge temperatures, therefore, lower gas turbine output power.

Combustor: Adds heat energy to the airflow. The output power of the gas turbine is directly proportional to the combustor firing temperature, i.e., the combustor is designed to increase the air temperature up to the material limits of the gas turbine while maintaining a reasonable pressure drop.

Gas Producer Turbine: Expands the air and absorbs just enough energy from the flow to drive the compressor. The higher the gas producer discharge temperature and pressure, the more energy is available to drive the power turbine, therefore, creating shaft work.

Power Turbine: Converts the remaining flow energy from the gas producer into useful shaft output work. The higher the temperature difference across the power turbine, the more shaft output power is available.

Exhaust System: Directs exhaust flow away from the gas turbine inlet. Often, a silencer is part of the exhaust system. Similar to the inlet system, the exhaust system is designed for minimum pressure losses. Figure 3 depicts these functions more clearly.

As seen above, the efficiency of a gas turbine depends on the turbine inlet temperature (TIT). Figure 4 shows a yearwise data on the steady rise in the turbine inlet temperature and its relationship with the efficiency of a gas turbine which also increases linearly with the TIT.[4]

Figure 5 further gives the same yearwise data of TIT but relates it to the material capability of the blade. As can be seen that with

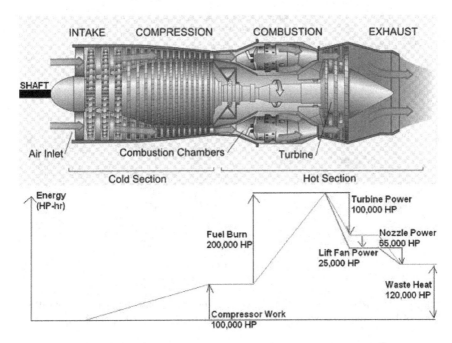

Fig. 3. Schematically shows energetics of various sections.[3]

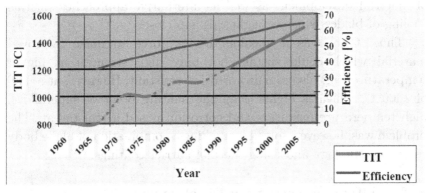

Fig. 4. The yearwise relationship between the TIT versus efficiency of a turbine.[4]

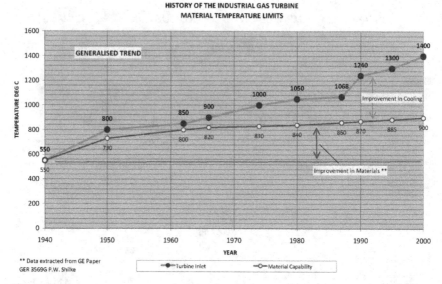

Fig. 5.   Yearwise rise in TIT and also the increasing gap between temperature limit of blade material with steady improvement in material capability.[5]

steady rise in TIT, the gap between the material temperature limit went on increasing, hence there was a continuous increase in the gap of material development as well as alternative approaches such as cooling of blades which brought this gap closer.[5]

Thus, there was a genuine driving force to make improved material which could sustain not only high strength at higher temperatures but also remain corrosion resistant. However, at a TIT of 1400°C, which is higher than the melting point of superalloys, high temperature oxidation and corrosion posed a real threat. The problem was, however, solved by making a combination of blade body cooling technology and using thermal barrier coatings.

2. High Temperature Alloy Development

Aircrafts can be classified as light, passenger, fighter or supersonic. The main variation is in terms of skin materials and gas turbine efficiencies. The skin material, in most of the airplanes, is aluminum alloy and also for the wings, fuselage, landing gear, etc., specific aluminum alloys are used. These airframes, and other components face only room temperature and low temperature corrosion problems, while materials in gas turbines are mainly superalloys, which in turn face severe high temperature corrosion problems.

Superalloys are austenitic materials, mainly Ni base, Fe–Ni or Co base with a host of substitutional solid solutioning strengthening elements, such as Mo, Ti, Ta, Nb, or W, grain boundary strengthening elements, such as Zr, Hf or B, grain refining elements, such as V, intermetallic precipitates, such as Ni_3Al, Ni_3Ti, or Ni_3Nb and various carbides, such as Mo_2C, TiC, VC, Cr_2C_3, which help in sustaining strength of the alloy at high temperatures. In addition, two elements, aluminum and chromium are the main alloying elements which help in providing oxidation resistance to superalloys by forming alumina or chromia protective layers.

Superalloy components can be made by three routes: wrought route, powder metallurgy or investment castings. Though wrought route gives the simplest route to fabricate component, it requires several steps involving, rolling, forging, extrusion with milling and surface cleaning at various stages. This results in a lot of wastage of material. The other problem is that many wrought alloys achieve high strength with several substitutional alloying elements. This makes a wrought superalloy of low incipient melting temperature and hence their lower utility in aerospace application. This also limits the heat treatment of the highly alloyed matrix with lots of precipitates and intermetallic phases. Segregation of high melting elements such as W, Mo, Hf, etc. poses another problem due to which the alloy results in poor mechanical properties.

The other method of fabrication is powder metallurgy which gives superior alloys with possibility to get near net shaped material with least wastage of starting materials. Powder-processing routes have been developed to overcome the difficulties associated with melt-related defects and are viable for the production of advanced high-strength polycrystalline superalloy components. Some of these advantages are as follows:

- To increase the strength of polycrystalline Ni-based superalloys, levels of refractory alloying additions and γ-forming elements have gradually increased to levels that make conventional processing routes deficient.

- Limited ductility of the conventional high-strength superalloys renders the ingot susceptible to cracking as thermally induced stresses evolve during cooling.

- Powder processing begins with gas or vacuum atomization of a highly alloyed vacuum induction melting (VIM) ingot. Rapid solidification of the fine powders effectively suppresses macro-segregation within the alloy.

- Because the low ductility associated with the corresponding high strength causes many of these advanced superalloys to be very sensitive to initial flaw sizes, the atomized powders are separated based on particle size.

- Powder sizes directly influence the initial potential crack size present in the finished component.

- Once powders are collected into steel cans, the cans are evacuated under vacuum and sealed. The cans are then hot isostatically pressed (HIP) or extruded to consolidate the powder. The HIP process consists of heating the alloy to just below the γ solvus temperature under a hydrostatic pressure of up to 310 MPa.

The investment casting is perhaps the best route to fabricate superalloys with excellent mechanical properties as well as with higher incipient melting temperature.[6] The conventional casting results in equiaxed grains due to cooling of the cast from all sides of walls. The biggest limitation of equiaxed cast alloys is the threat of failure of grains which are perpendicular to the stress axis. Thus, an alternate to equiaxed grain structure is aligned grains in the direction of stress axis. Such investment castings are called as directionally solidified castings and are found to be more safer than the former. In order to make the alloy more safer, it was postulated to have an alloy with no grain boundaries, that is a single crystalline alloy. There are more than one advantages of a single crystalline alloy.

Since there are no grains, there is no requirement of grain boundary strengthening elements such as B, Zr, or Hf. Also, element such as V which controls the grain size is not required. Thus, removing these elements results in higher incipient melting point of the single crystalline alloy. This helps in carrying out homogenization treatment at still higher temperature, thus avoiding any segregation of heavy elements and providing better mechanical properties.

The fabrication of directionally solidified and single crystalline cast alloys requires monitoring of two important parameters. These are rate of unidirectional cooling and the temperature gradient between solidifying front and the melt. Their relationship is shown in Fig. 6 which shows that between directionally solidified (DS) cast alloys and single crystal (SC) cast alloy, the main difference is in their thermal gradient and cooling rate. While DS cast alloy requires lower thermal gradient (32°C) and lower cooling rate, SC cast alloy requires higher thermal gradient (32–72°C) and much higher cooling rates.[7]

Today, most of the turbine blade materials in many passenger and fighter planes are made of single crystal superalloys. The compositions of various superalloys, prepared by different routes, are given in Table 1.[8] Figure 7 compares the grain structures of three superalloys prepared by three different casting methods.[9]

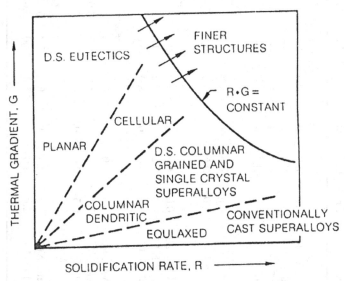

Fig. 6. Relationship between thermal gradient between hot liquid and solidification front versus unidirectional cooling rate.[7]

Table 1: Composition of various superalloys.[8]

Alloy	Cr	Co	Mo	W	Ta	Re	Nb	Al	Ti	Hf	C	B	Y	Zr	Other
Conventionally Cast Alloys															
Mar-M246	8.3	10.0	0.7	10.0	3.0	—	—	5.5	1.0	1.50	0.14	0.02	—	0.05	—
Rene' 80	14.0	9.5	4.0	4.0	—	—	—	3.0	5.0	—	0.17	0.02	—	0.03	—
IN-713LC	12.0	—	4.5	—	—	—	2.0	5.9	0.6	—	0.05	0.01	—	0.10	—
CI023	15.5	10.0	8.5	—	—	—	—	4.2	3.6	—	0.16	0.01	—	—	—
Directionally Solidified Alloys															
IN792	12.6	9.0	1.9	4.3	4.3	—	—	3.4	4.0	1.00	0.09	0.02	—	0.06	—
GTD111	14.0	9.5	1.5	3.8	2.8	—	—	3.0	4.9	—	0.10	0.01	—	—	—
First-Generation Single-Crystal Alloys															
PWA 1480	10.0	5.0	—	4.0	12.0	—	—	5.0	1.5	—	—	—	—	—	—
Rene' N4	9.8	7.5	1.5	6.0	4.8	—	0.5	4.2	3.5	0.15	0.05	0.00	—	—	—
CMSX-3	8.0	5.0	0.6	8.0	6.0	—	—	5.6	1.0	0.10	—	—	—	—	—
Second-Generation Single-Crystal Alloys															
PWA 1484	5.0	10.0	2.0	6.0	9.0	3.0	—	5.6	—	0.10	—	—	—	—	—
Rene' N5	7.0	7.5	1.5	5.0	6.5	3.0	—	6.2	—	0.15	0.05	0.00	0.01	—	—
CMSX-4	6.5	9.0	0.6	6.0	6.5	3.0	—	5.6	1.0	0.10	—	—	—	—	—
Third-Generation Single-Crystal Alloys															
Rene' N6	4.2	12.5	1.4	6.0	7.2	5.4	—	5.8	—	0.15	0.05	0.00	0.01	—	—
CMSX-10	2.0	3.0	0.4	5.0	8.0	6.0	0.1	5.7	0.2	0.03	—	—	—	—	—

(Continued)

Table 1: (Continued)

Alloy	Cr	Co	Mo	W	Ta	Re	Nb	Al	Ti	Hf	C	B	Y	Zr	Other
Wrought Superalloys															
IN 718	19.0	—	3.0	—	—	—	5.1	0.5	0.9	—	—	0.02	—	—	18.5Fe
Rene' 41	19.0	11.0	10.0	—	—	—	—	1.5	3.1	—	0.09	0.005	—	—	—
Nimonic 80A	19.5	—	—	—	—	—	—	1.4	2.4	—	0.06	0.03	—	0.06	—
Waspaloy	19.5	13.5	4.3	—	—	—	—	1.3	3.0	—	0.08	0.006	—	—	—
Udimet 720	17.9	14.7	3.0	1.3	—	—	—	2.5	5.0	—	0.03	0.03	—	0.03	—
Powder-Processed Superalloys															
Rene' 95	13.0	8.0	3.5	3.5	—	—	3.5	3.5	2.5	—	0.065	0.013	—	0.05	—
Rene' 88 DT	16.0	13.0	4.0	4.0	—	—	0.07	2.1	3.7	—	0.03	0.015	—	—	—
N18	11.2	15.6	6.5	—	—	—	—	4.4	4.4	0.5	0.02	0.015	—	0.03	—
IN100	12.4	18.4	3.2	—	—	—	—	4.9	4.3	—	0.07	0.02	—	0.07	—

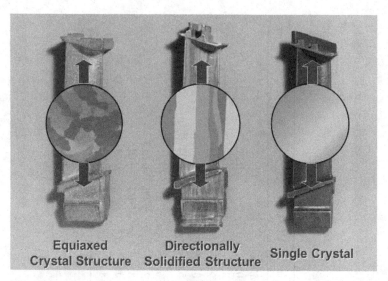

Equiaxed Crystal Structure   **Directionally Solidified Structure**   **Single Crystal**

Fig. 7. Grain structure of a conventionally solidified equiaxed, directionally solidified and single crystalline, superalloy along with representative microstructure.[9]

The single crystalline superalloys are further classified into generations 1–3, as given in Table 1. Generation 1 superalloys are conventional superalloys which are strengthened by one or more intermetallic precipitates, along with a host of substitutional solid solutioning elements, and carbide formers. Corrosion resistance is provided by either chromia or alumina protective oxide layers. To provide high corrosion resistance, a high concentration of Cr is required in the alloy. However, too much of Cr is usually not desirable as it affects the mechanical properties of the cast alloy by enhancing the coarsening of γ' precipitates. Modifying the composition by reducing Cr to half (5%), doubling Co, but especially adding rehenium (Re), helped in creating a superalloy which has better creep capability over 1st generation alloy by over 10°C.[10] The fine microstructure by a combination of three heat treatments gave a microstructure with excellent distribution of extremely tiny γ' as shown in Fig. 8. Enhancement in the creep resistance, UTS and yield strength and hot corrosion behavior also took place as shown in Fig. 9.

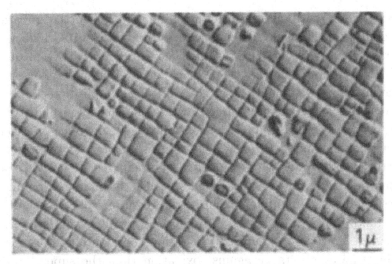

Fig. 8. Fully heat-treated (1340°C 4 hours + 1080°C 4 hours + 700°C 24 hours) microstructure of PWA 1484 2nd generation single crystalline alloy showing uniform distribution of tiny γ' precipitates.[10]

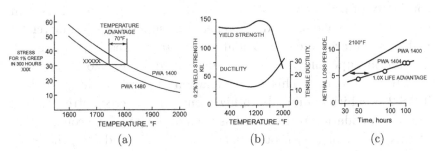

Fig. 9. Comparison of 1st generation PWA single crystal PWA 1480 versus 2nd generation single crystal alloy PWA 1484: (a) improved creep resistance, (b) improved tensile properties, and (c) improved oxidation resistances.[10]

In the third generation, single crystalline superalloys were further modified by increasing the Re level to 5%–6%, reducing Cr still further and increasing Co. This resulted in still superior properties. Quest for improving properties of superalloys increased further and 4th and 5th generation superalloys were developed.[11] 4th generation Ni-base superalloys contain 2–3 wt.% Ru, which hinders the precipitation of topologically close packed (TCP) phases[12] and improves the high temperature microstructure stability.[13-15] Fourth generation superalloys have achieved temperature capabilities 30°C higher on average than those of the previous generation superalloys in terms of high temperature creep strength. 5th generation superalloys have been invented by the optimization of alloying compositions, and the content of Ru has increased to 5–6 wt.%; the lattice misfit between the γ and the γ' phases has been controlled to balance the interfacial strengthening and coherency, and the dislocation network at the interface of the γ and the γ' phases has become finer than that of 4th generation superalloys in order to inhibit dislocation migration under stress. Thus, the high temperature creep resistance of 5th generation superalloys is better than 4th generation superalloys.[16] However, it appeared that the 4th and 5th generations of superalloys are likely to have lower resistance against oxidation than the superalloys of 2nd and 3rd generations owing to the higher content of refractory elements such as Mo, Re and Ru.[17] These refractory-based oxide species have relatively higher vapor pressures and can disrupt the continuity of protective Al_2O_3 formed on the surface during thermal exposure. To make 4th and 5th generation superalloys commercially viable, an improvement in oxidation resistance was imperative. 6th generation superalloys were then developed by Kyoko Kawagishi1 in NRIM, Japan.[11] This 6th generation superalloy designated as TMS-238, which exhibits both high temperature creep strength and improved oxidation resistance, has been developed. The composition of this alloy along with other alloys of 3rd, 4th, and 5th generation alloys is given in Table 1 and their mechanical properties in Tables 2 and 3, respectively.

As predicted, these alloys showed superior oxidation and hot corrosion behavior as shown in Figs. 10 and 11. Figure 9 shows improved oxidation behavior in cyclic oxidation tests. 6th generation TMS 238 shows almost no spalling compared to 5th generation TMS 196 or 4th generation PEA1497. In the same way, the mass loss in hot corrosion tests is compared in Fig. 10.

Finally, Fig. 12 summarizes the overall comparison on oxidation behavior and creep rupture life of 3rd, 4th, 5th, and 6th generation

Table 2: Chemical composition of 3rd–6th generation single crystals super alloys.[11]

Alloy	Co	Cr	Mo	W	Al	Ti	Ta	Hf	Re	Ru
CMSX-4	9.6	6.4	0.6	6.4	5.6	1.0	6.5	0.1	3.0	0.0
MX-4/PWA1497	16.5	2.0	2.0	6.0	5.55	0.0	8.25	0.15	5.95	3.0
TMS-138A	5.8	3.2	2.8	5.6	5.7	0.0	5.6	0.1	5.8	3.6
TMS-196	5.6	4.6	2.4	5.0	5.6	0.0	5.6	0.1	6.4	5.0
TMS-238	6.5	4.6	1.1	4.0	5.9	0.0	7.6	0.1	6.4	5.0

Table 3: 0.2% Yield strength and UTS at two temperatures for 3rd–6th generation superalloys.[11]

	400°C		750°C	
Alloy	0.2% yield	UTS	0.2% yield	UTS
CMSX-4	860	950	950	1150
TMS-138A	830	906	868	1241
TMS-196	879	1214	845	1308
TMS-23S	925	1373	1041	1348

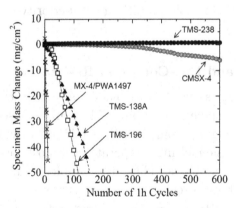

Fig. 10. Cyclic oxidation test behavior of 6th generation superalloys TMS-238 versus 3rd–5th generation superalloys. As can be seen, TMS-238 shows the maximum resistance to spalling.[11]

A.S. Khanna & V.S. Agarwala

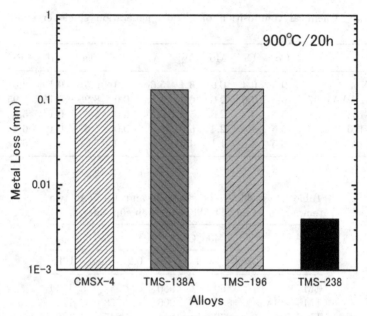

Fig. 11.   Hot corrosion behavior of 3rd–6th generation superalloys, tested in 75% Na2SO4+25% NaCl at 900°C for 20 hours.[11]

alloys in Fig. 12 which shows ultimate superiority of TMS-238 over other alloys.

3. Oxidation and Hot-Corrosion Resistance

Components, such as vanes and blades, are the most critical components of an airplane gas turbine. They are exposed to very high temperatures, ranging from 1100°C to 1500°C, which is actually above or close to the melting temperature of most corrosion resistant alloys such as superalloys. Thus, the problems such as oxidation, sulfidation and hot corrosion are severe threats. In addition, creep, fatigue and strength stability is another serious problem of these alloys. Erosion due to foreign particles can also be a big threat. A close look into literature gives a list of number of failures of blades due to one or more of these problems.[18–21] As discussed in the above section, superalloys are the most suitable materials for turbine component application. Figure 13 summarizes the list of various

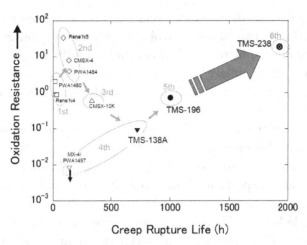

Fig. 12. Summary of the oxidation behavior and creep rupture behavior of 2nd–6th generation single crystal alloys.[11]

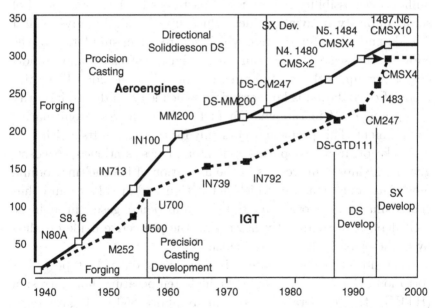

Fig. 13. Progress in aero and industrial blades metallurgy with time.[22]

superalloys with its development year and its corresponding use for a particular component of a gas turbine.[22] Beyond the year 2000, there was further development of 2nd generation and 3rd generation single crystal superalloys, where part of Cr is replaced by Co and rhenium and more concentration of refractory metals as substitutional solid solutioning elements.

From the corrosion resistance point of view, superalloys can be either alumina former or chromia former. Chromia former alloys limit their utility up to 900°C because of destabilization of chromia layer after this temperature. The protective chromia layer changes to $Cr(VI)$ oxide which is volatile and hence destabilizes the protective $Cr(III)$ oxide layer.[23] For applications above 900°C, mainly alumina forming superalloys are used. Material degradation of superalloys in gas turbines can be due to oxidation, sulfidation and hot corrosion. Oxidation environment is obvious because of presence of air. Sulfidation possibility is because of parts per million (ppm) level of sulfur impurity in aviation fuel, which upon oxidation gives sulfur dioxide and/or sulfur trioxide which can result in sulfidation of the alloy or the scale formed can be a mixture of oxide and sulfide, depending upon the partial pressure of sulfur dioxide. The third mode of degradation is hot corrosion, which is again due to impurity of Mo and/or V in the aviation fuel and salt ingress from marine environment. This problem occurs only for those aircrafts which land and take off from sea ports, or fighter planes, stationed on coast-guard sea ships. This results in the formation of low melting sodium moybdate and vanadate, which deposit on turbine blades and thus cause enhanced corrosion due to faster diffusion of gases through the melt deposit or by directly destroying the protective oxide on alloy by acidic or basic fluxing mechanism.[23]

The hot corrosion is categorized in two groups. In type 1, hot corrosion which is designated as high temperature hot corrosion (HTHC), the temperatures are usually above 850°C. Figure 14(a) shows the failure of turbine blade due to type 1 hot corrosion. Low temperature hot corrosion (LTHC), where the corrosion process occurs at temperature in the range 700–800°C, which is lower than the melting temperature of Na_2SO_4. This type of hot corrosion is

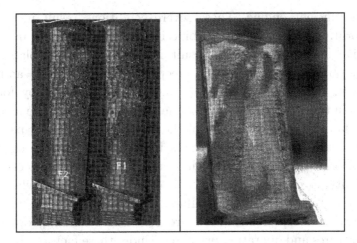

Fig. 14. Showing the failure of blades due to hot corrosion.[20]

also called type 2 hot corrosion. Here, the deposit melts only after impurities such as V or Mo react with sodium sulfate to form low melting sodium vanadate or molybdate. Vanadium is an unavoidable contaminant in certain fuels. When vanadium containing deposits are exposed to high temperatures, accelerated corrosion takes place, resulting in the formation of a highly aggressive liquid phase at temperature as low as 535°C, depending upon the ratio of Na/V. They also enhance the solubility of protective oxide in combination with Na_2SO_4. This type of corrosion is shown in Fig. 14(b).[20]

4. Failure Modes of a Turbine Blade

The main function of a turbine blade in the high pressure turbine is to extract power and energy from the hot gas stream, entering from the engine combustion chamber to drive the upstream compressor section of the engine. These blades operate at speeds over 10,000 rpm and extract up to 750 horse power of energy. They have on average about 15,000 hours of service life. These blades work in the hottest part of the engine core in gas temperatures, reaching as high at 1595°C.[8] The gas temperature entering the turbine, called turbine inlet temperature (TIT), is directly related to the overall engine efficiency. Higher TIT results in more power, thus providing a more efficient engine. These severe conditions result in various deterioration modes for gas turbine blades. Factors that can create corrosive environments, and result in deterioration of blades are foreign object damage (FOD), creep, fatigue, oxidation, and hot corrosion. Figure 14 shows dents and scratches caused by foreign objects in the air stream of the engine. These objects can be ingested by the fan at the inlet section of the engine or created by particles of metal and hard carbon deposits from the combustion chamber section. These dents and scratches created by foreign objects have the ability to create stress concentrations in the blade, which can at a later stage cause fatigue failure. The three largest factors that can lead to creep deformation are centrifugal loading, operating temperatures and operating pressures. Under these factors, the blade material is weakened resulting in the acceleration of the creep process. This process can eventually form voids that will lead to the initiation of small cracks in the blade which can then grow in size and eventually lead to blade failures. These failures can be as small as minor cracks or as devastating as a blade cracking right in half. These failures are typically known as stress rupture. Another issue that can arise from creep is the dimensional changes that can occur, reducing its aerodynamic efficiency and even leading to elongation and rubbing of the blade tip, which in turn can lead to vibration and noise.[9] Figure 16 shows a blade damage by creep.

Another important degradation mode is fatigue in which the deterioration occurs by crack initiation and growth, resulting from

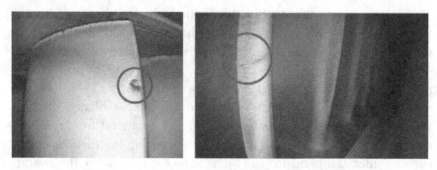

Fig. 15.   Damage of turbine due to foreign objects.[9]

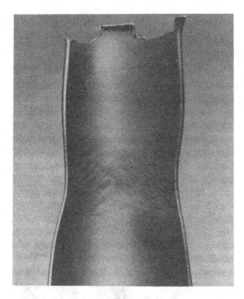

Fig. 16.   Creep damage of turbine blade.[9]

the creation of alternating stresses during engine operation by either of three distinct modes of alternating stresses, low cycle fatigue (LCF), high cycle fatigue (HCF) or thermal mechanical fatigue (TMF). LCF is typically created with the alternating of stresses developed in the rotating blades through the change in rotational speeds during engine operation, in particular the engine starting up and the engine shutting down. LCF cracks are typically initiated in regions where the developed stresses through rotation are the highest; these are areas like the rims of rotors, turbine disks and blade to disk connections.[9,20] This mode is typically characterized by high amplitude low frequency plastic strains. LCF is typically prevented by the retiring of parts after a specified number of loading cycles. HCF is typically generated due to the alternating stresses created in turbine blades through the excitement of a resonant state through variations in gas impulse loads (Fig. 16). This mode is usually characterized by low amplitude, high frequency elastic strains. Two main abnormal conditions lead to the creation of these types of variation, the first being the creation of abnormalities in the

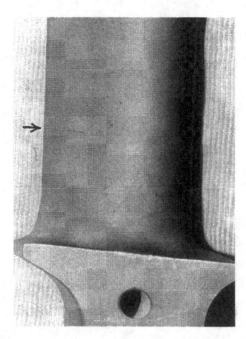

Fig. 17. Failure of turbine blade due to fatigue.[20]

combustion process while the second is the creation of gas velocity differences through abnormalities in the shape of individual turbine vanes upstream of the turbine blades. HCF is typically prevented by maintenance of uniform combustion and uniform shape of the turbine vanes. TMF is typically generated by the alternating stresses, created by change in gas temperatures, resulting in different hot and cold regions of the turbine blade. This mode of fatigue tends to occur in regions that are the thinnest in cross-section and where the greatest thermal strains are located. In turbine blades, a rapid reduction in gas temperatures can result in a thin trailing edge of the blade cooling faster than a thicker mid-span and leading edge region. This difference in cooling rates can result in a difference in thermal contraction between the trailing edge and the rest of the blade, resulting in the creation of stresses at the trailing edge region. TMF can also lead to its own form of cracking when compared to LCF and HCF. TMF is usually prevented by retiring the blade after

Fig. 18.　Failure of turbine blades and vanes by hot corrosion and fatigue.[20]

it has reached a set number of cycles and also limiting the severity of thermal strains. A turbine rotor from an aero-derivative gas turbine that experienced a fatigue failure combined with hot corrosion is shown in Fig. 18.[20]

5. Protection Technology of Turbine Blades and Vanes

As discussed above, the gas turbine components such as blades and vanes are usually made by highly sophisticated superalloys, fabricated by the most advanced technique of single crystal formation. However, inspite of making such an advanced alloy, it still has several limitations to be used for turbine components. This is mainly because the alloy is exposed to an environment which has much higher temperature than the melting point of the alloy. As discussed above, the single crystalline superalloy of even 6th generation has an M.P. of 1320–1330°C, which is a few hundred degrees lower than the gas turbine temperature. Thus, the blade material needs to be modified so that it can experience a temperature lower than 1300°C or so.

Fig. 19.   Design of a turbine blade with hole geometry to cool the inner surface.[9]

This can be achieved by two approaches. One by cooling the blade by creating a special geometry of holes from where cool gas can be passed which can help in bringing the temperature of the blade lower. A typical turbine blade with several holes is shown in Fig. 19. The second approach is to introduce thermal barrier coatings. A thermal barrier coating, as the name suggests, insulates the blade material. The temperature drop can be as large as 150°C per 100 μm thickness. The best thermal barrier material can be a ceramic. Table 4 lists various ceramics with their physical and thermal properties. Out of this, ZrO_2 has been found to be the best choice for thermal barrier coatings. Further, in order to meet the mechanical properties of ZrO_2, it is necessary to add about 8% of yttria, which gives it good rupture properties as shown in Fig. 20. It shows that addition of about 8% of yttria stabilizes ZrO_2 in such a manner that its failure is delayed to the highest number of cycles.[24]

The next problem is that the ceramic coatings cannot be applied directly on the alloy surface because of two reasons: (i) large difference in the thermal expansion of superalloy and the ceramic,

Table 4: Physical properties of various ceramics.

Property	HfO_2	ZrO_2	Y_2O_3	Al_2O_3	SiO_2
Melting point, °C	2810	2700	2460	2015	1728
Density, g/cm^3	9.68	5.60	5.03	3.98	2.32
Thermal expansion, ppm/°C	6.8	7.5	6.8	8.1	0.5
Limit of stability with carbon, °C	1700	1600	—	1900	—
Vapor pressure, Pa					
$\times 10^{-6}$ at 1650°C	0.13	10.6	—	133	—
$\times 10^{-5}$ at 1927°C	3.93	127	—	50	—
$\times 10^{-3}$ at 2200°C	3.26	78.7	—	20	—
Evaporation rate, μm/hr $\times 10^{-5}$					
at 1650°C	6.7	670	—	—	—
at 1927°C	0.019	0.75	—	—	—
at 2200°C	1.4	44	—	—	—
Crystal strucure	FCC	FCC	BCC	HCP	CT
Oxygen permeability, g/cm · sec					
$\times 10^{-13}$ at 1000°C	360	120	9.5	—	—
$\times 10^{-11}$ at 1400°C	72	37	2.5	—	—
$\times 10^{-10}$ at 1800°C	46	30	2.0	—	—

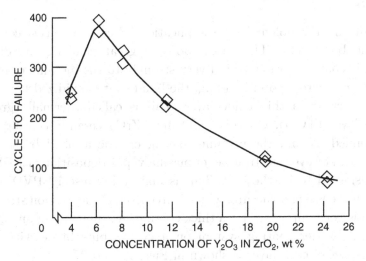

Fig. 20. Cycles to failure of ZrO_2 as a function of yttria addition in ZrO_2.[24]

and (ii) ZrO_2 coating is not a corrosion resistant coating. Hence before applying the ceramic coating, a corrosion resistant coating needs too be applied first. Though there is a history of large number of coatings being applied on superalloys that can be seen in Refs. 25 and 26, the most acceptable among many for oxidation and hot corrosion resistance are called MCrAlY coatings, where "M" stands for base metal of the alloy, whether Ni, Fe–Ni, or Co, based upon whether the alloy base is Ni, Fe–Ni, or Co, respectively. Cr is essential element for oxidation and sulfidation and hot corrosion resistance, aluminum takes care of alumina oxide formation and Y is added for enhancing the oxide adherence. These corrosion resistant coatings are also called bond coat. The main purpose of using these coatings over pure aluminide or modified aluminide coating was their better thermal expansion match with base alloy and more corrosion resistance. This coating can be applied by several methods. Plasma spray has been considered as the most acceptable as it gives a clean homogenous coating with good adherence and reasonably low porosity. Other methods such as HVOF and Cold Spray can also be used.[27]

The next problem is the application of ceramic ZrO_2 coating on the bond coat. This again poses a great challenge as direct ceramic coating does not bind very strongly to the metallic coating. Hence to create a good bonding, the bond coat is oxidized for some time to create a thin oxide layer. This is called thermally grown oxide layer (TGO). On this TGO, now ZrO_2 ceramic coating can be applied. Again, the ceramic coating can be applied by several methods, however electron beam plasma vapor deposition (EBPVD) is considered to be the best. This is mainly because EBPVD gives columnar coating structure, which is free from all deposition stresses, which is very important for a thick coating. A schematic of an actual thermal barrier coating system on a gas turbine blade with real thicknesses of each layer is shown in Figs. 21 and 22.

Figure 23 shows the schematic of the thermal barrier coating showing the columnar structure of the outer ceramic coating. This kind of structure is stress free and remains stable under thermal cycling conditions.

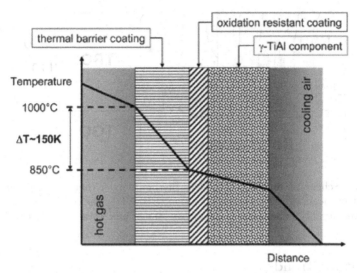

Fig. 21. Schematic of a thermal barrier coating applied to γ-TiAl alloys. The technology is adopted from nickel-based superalloys used in gas turbines.[28]

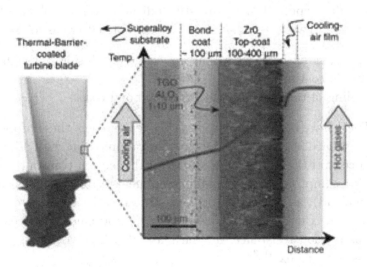

Fig. 22. Actual coating on a turbine blade with thickness of various layers.[29]

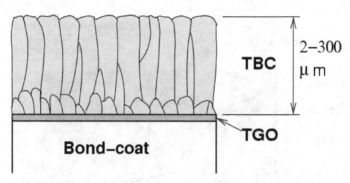

Fig. 23.   Schematic microstructure of a thermal barrier coating (TBC) obtained by electron beam physical vapor deposition (EBPVD). The columnar microstructure considerably enhances the strain resistance and therefore thermal cycling life.[30]

6. Future Trends

Though thermal barrier coatings in combination with air cooling of blades has proved to be a suitable technology to run gas turbines successfully, reported literature still indicate several problems from this system. Two most common problems noticed are as follows:

1. Oxidation of bond coat due to oxygen diffusion through the ceramic coating ZrO_2.
2. Spalling of the ceramic coat from the bond coat.

There have been several approaches to overcome these problems. One approach is to modify the zirconia ceramic coat with additives which control the spallation and control the oxygen transport. More complex compositions (e.g., multicomponent stabilization based Y, Yb, and Gd), as well as pyrochlore structures based on zirconates $(La_2Zr_2O_7, Gd_2Zr_2O_7)$ to meet new performance requirements such as erosion resistance, molten glass attack (CMAS) as well as increased heat flux.[31–33]

Another approach to overcome spalling problem was to make thermal gradient coatings. For example, instead of a 100% ceramic coating on bond coat, a series of coatings with increased level of ceramic component ZrO_2 was added. This would create better compatibility of various layers in terms of thermal expansion mismatch

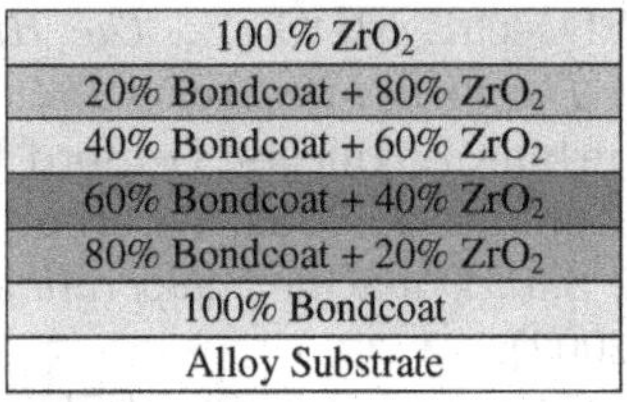

100 % ZrO₂
20% Bondcoat + 80% ZrO₂
40% Bondcoat + 60% ZrO₂
60% Bondcoat + 40% ZrO₂
80% Bondcoat + 20% ZrO₂
100% Bondcoat
Alloy Substrate

Fig. 24.   Schematic of a gradient thermal barrier coating.

and hence restrict spalling. A schematic of such a thermal gradient coating is shown in Fig. 24 where the first layer on metal is a complete bond coat, with subsequent layers, varying in the ZrO_2 contact. Lots of work has been done by several workers on this concept.[34–36]

Khor *et al.*[34] reported functionally gradient coatings of NiCoCrAlY with varying percentages of yttria stabilized Zirconia using plasma spray technique. Composite samples were made with powders of NiCoCrAlY and 8%Y2O3 stabilized zirconia with various layers with varying amounts of YSZ from 25% to 100%. The five layered functionally gradient coating showed attractive properties compared to the duplex TBC coatings. The new microstructure of the coating resulted in reduced residual stresses and hence better stability of coating in terms of spalling resistance.

Though the concepts look great, its commercial application on turbine blades has still not progressed. The research is still continuing and perhaps there would be a gradient coating on blades one day.

References

1. V.S. Agarwala, Aircraft Corrosion Control and Maintenance, RTO-MP-AVT-144,   *ftp.rta.nato.int/public//PubFullText/RTO/MP/...///MP-AVT-144-25.pdf.*
2. K. Steddenbenz, *https://t-quare.gatech.edu/access/content/group/...//Gas%20Turbines.ppt.*
3. *http://www.comsol.com/blogs/turbine-stator-blade-cooling-and-aircraft-engines/.*
4. M.R. Susta and M. Greth, Efficiency improvement possibilities in CCGT power plant technology, IMTE AG, Power Generation Consulting Engineers, Switzerland, *www.mrsphoto.net/4-IMTE%20AG/2-PGA-2001.pdf.*

5. R.J. Hunt, CENG FIMechE FIDGTE, *The History of the Industrial Gas Turbine (Part 1, The First Fifty Years 1940–1990)*, The Institution of Diesel and Gas Turbine Engineers, Bedford Heights, Manton Lane, Bedford MK41 7PH, publication 582, Peterborough, 2011.

6. M.J. Donachie and S.J. Donachie, *Superalloys: A Technical guide*, (ASM Publication, 2002).

7. F.L. VerSynder, Superalloy technology — today and tomorrow, in *Proc. High Temperature Alloys for Gas Turbine Application* by R. Brunetad, D. Coutsouradis, T.B. Gibbons, Y. Lindblom, D.B. Meadowcro and R. Sticker (Eds.) (Reidel Publishing, 1982), pp. 1–49.

8. T.M. Pollock and S. Tin, Nickel-based superalloys for advanced turbine engines: chemistry, microstructure, and properties, *J. Propulsion Power*, **22**, No. 2, (2006), pp. 361–374.

9. D.C. Nadeau, The development of a thermal model to predict corrosion propensity on internal features of turbine airfoils, Master of Engineering in Mechanical Engineering, Rensselaer Polytechnic Institute, Hartford, Connecticut, USA, January, 2013.

10. A.D. Cetel and D.N. Duhl, Second-generation nickel-base single crystal superalloy, in superalloys 1988, S. Reichman, D.N. Duhl, G. Maurer, S. Antolovich and C. Lund (Eds.), (The Metallurgical Society, 1988), pp. 235–244.

11. K. Kawagishi, A.-C. Yeh, T. Yokokawa, T. Kobayashi, Y. Koizumi and H. Harada, Development of an oxidation-resistant high-strength sixth generation single-crystal superalloy tms-238, in Superalloys 2012, 12th International Symposium on Superalloys, E.S. Huron, R.C. Reed, M.C. Hardy, M.J. Mills, R.E. Montero, P.D. Portella and J. Telesman (Eds.), (TMS, The Mineral, Metals and Materials Society, 2012), pp. 189–196.

12. K. O'Hara, W.S. Walston, E.W. Ross and R. Darolia, U.S. Patent 5,482,789, Nickel base superalloy and article, 1996.

13. Y. Koizumi, T. Kobayashi, T. Yokokawa, H. Harada, Y. Aoki, M. Arai, S. Masaki and K. Chikugo, Development of 4th generation single crystal superalloys, in *Proc. 2nd International Symposium on High Temperature Materials*, Tsukuba, Japan, 2001, pp. 30–31.

14. J.X. Zhang, T. Murakumo, Y. Koizumi, T. Kobayashi, H. Harada and S. Masaki, Interfacial Dislocation Networks Strengthening a Fourth-Generation Single-Crystal TMS-138 Superalloy, *Metall. Mater. Trans. A*, **33** (2002), pp. 3741–3746.

15. Y. Koizumi, T. Kobayashi, T. Yokokawa, J.X. Zhang, M. Osawa, H. Harada, Y. Aoki and M. Arai, Development of Next-Generation Ni-Base Single Crystal Superalloys, *Superalloys 2004*, (TMS, 2004), pp. 35–43.

16. S. Walston, A. Cetel, R. MacKay, K. O'Hara, D. Duhl and R. Dreshfield, Joint development of a fourth generation single crystal superalloy, *Superalloys 2004*, (TMS, 2004), pp. 15–24.

17. K. Kawagishi, A. Sato, T. Kobayashi and H. Harada, Effect of alloying elements on the oxidation resistance of 4th generation Ni-base single-crystal superalloys, *J. Japan Inst. Metals* **69** (2005), pp. 249–252.

18. N. Eliaz, G. Shemesh and R.M. Latanision, Hot Corrosion in Gas Turbine Components, Engineering Failure Analysis, *Eng. Fail. Anal.* **9** (2002), pp. 31–43.

19. Hot Corrosion in Gas Turbines, Chapter 9, High Temperature Corrosion and Materials Application (#05208G), http://www,asminter national.org.,

20. C.B. Meher-Homji and G. Gabriles, Gas turbine blade failures-causes, avoidance, and troubleshooting, in *Proc. 27th Turbomachinery Symposium*.

21. M.R. Orlov, High-temperature corrosive damage to superalloys during operation of blades of gas-turbine engines and power plants, *Polymer Science Series D*, **6**, (2013), http://link.springer.com/journal 12260/6/3/page/1. pp. 250–255.

22. I. Gurrappa, I.V.S. Yashwanth, I. Mounika, H. Murakami and S. Kuroda, in *The Importance of Hot Corrosion and Its Effective Prevention for Enhanced Efficiency of Gas Turbines*, I. Gurrappa (Ed.) (2015), http://www.intechopen.com/books/gas-turbines-materials-modeling-and-performance.

23. A.S. Khanna, in *Introduction to High Temperature Oxidation and Corrosion* (ASM International, 2008).

24. A. Miller, Thermal Barrier Coatings for Aircraft Engines, *J. Thermal Spray Technology*, **6**(1), (1997), pp. 35–42 (ASM International, 1997).

25. W.J. Quadakkers, A.K. Tyagi, D. Clemens, R. Anton, and L. Singheiser, The significance of bond coat oxidation for the life of TBC coatings, in *Elevated Temperature Coatings: Science and Technology*, J.M. Hampikian and N.B. Dahotre (Eds.) (TMS, Warrendale, PA, 1999).

26. Amit Pandey, Vladimir K. Tolpygo, Kevin J. Hemker, Thermomechanical behavior of developmental thermal barrier coating bond coats, *J. Metals* **65**(4), (2013), pp. 542–549.

27. A.S. Khanna and W.S. Rathod, Development of CoNiCrAlY oxidation resistant hard coating using HVOF and cold spray techniques, *Int. J. Refract. Met. H.* **49** (2015), pp. 374–382.

28. C. Leyens, R. Braun, M. Fröhlich, and P. Eh. Hovsepian, Recent Progress in the Coating Protection of Gamma Titanium-Aluminides, *J. Metals* **58**(1) (2006), pp. 17–21.

29. N. Padture *et al.*, *Science* **296** (2002), p. 280.
30. K.A. Marino *et al.*, *PNAS* **108** (2011), pp. 5480–5487.
31. D. Zhu and R. Miller, *Int. J. Appl. Ceram. Technol.* **1** (2004), p. 86.
32. R. Vassen, X. Cao, F. Tietz, D. Basu and D. Stöver, *J. Am. Ceram. Soc.* **83**(8) (2000), p. 2023.
33. M. Jarligo, D.E. Mack, R. Vassen and D. Stöver, *J. Therm. Spray Technol.* **18**(2) (2009), p. 187.
34. K.A. Khor, Z.L. Dong, Y.W. Gu, Plasma sprayed functionally graded thermal barrier coatings, *Mater. Lett.* **38** (1999), pp. 437–444. Retrieved from   http://citations.springer.com/item?doi=10.1007/s11837-013-0551-1 "\t"_blank.
35. A. Portinha, V. Teixeira, J. Carneiro, J. Martinsb, M.F. Costa, R. Vassen and D. Stoever, Characterization of thermal barrier coatings with a gradient in porosity, *Surf. Coat. Technol.* **195** (2005), pp. 245–251.
36. D.R. Clarke, M. Oechsner and N.P. Padture, Thermal-barrier coatings for more efficient gas-turbine engines, *MRS Bulletin* **37**(10), (2012).

Chapter 7

Coatings for High Temperature Applications

N. I. Jamnapara* and S. Mukherjee

FCIPT, Institute for Plasma Research
Gandhinagar 382016, India
**nirav@ipr.res.in*

Coatings that survive high temperature and provide compatibility with the environment as well as with the substrates play an important role in different industrial sectors such as manufacturing, energy sector, transportation, waste-remediation, nuclear, space etc. Based on the type of application, the coatings have been developed so as to combat corrosion, wear, heat transfer, mass transport of unwanted species from environment into the bulk etc. Ceramics, intermetallics, cermets etc., have been known to possess high temperature stability. Additionally, some coatings demand functional properties along with high temperature stability, such as in solar applications. The present chapter discusses about different high temperature applications specific to energy sector such as thermal power, fusion reactors, fission reactors, solar power plants where coatings play a critical role and the factors affecting their selection and processing step have been elaborated.

1. Introduction

High temperature processing is gaining importance in different sectors of industry such as manufacturing, power generation, transportation, space, defence, environment (waste remediation) etc. Specifically, high temperature operation is of vital importance in energy sector where the difference in temperature (ΔT) decides the efficiency of the power system. In other words, the higher the operating temperature of power plants, the better the efficiency of the power plant. For example, the power and fuel economy demonstrated by modern gas turbines is strongly dependent upon and is limited by the high temperature strength of superalloys used in the hottest sections. Materials that are capable to operate at temperatures close to their melting points are crucial and referred to as superalloys. Conventionally, the structural materials such as steels can be used only till $0.3\,T_m$ (T_m = melting temperature of alloy). Bulk materials such as Ni-based, Fe–Ni-based and Co-based superalloys have been developed over the last four decades and can operate upto $0.8\,T_m$ temperature. At high temperatures, one of the most limiting factors of such superalloys is their ability to operate under severe environments. The interaction of the bulk or structural material with the environment leads to degradation due to surface dominant phenomena such as oxidation, corrosion, wear etc. As a result, the mechanical property of structural material gets modified with change in environment.[1]

With the above listed surface related issues, it becomes inevitable to develop coatings which can combat the environmental conditions while being intact with the substrate. Coatings are not just important to protect the substrate, but to enhance the service life of components and thereby improve the reliability of the total engineering assembly.

The subsequent chapter gives an insight into the challenges in high temperature operations, type of coatings and processes and applications of coatings used at high temperatures for different industrial sectors.

2. Issues at High Temperature

High temperature exposure not only reduces the bulk material strength and mechanical properties, but also accelerates the reaction of surface with the environment. Depending on the type of environment and the service conditions, a number of issues are generated which are undesirable for the reliable operation of the machines or systems. Some of the common issues associated with the surface dominant material degradation are briefly summarized in the following sections.

2.1 *Oxidation*

Metals or alloys oxidize when subjected to high temperature in air or highly oxidizing conditions such as combustion atmospheres with air or excess oxygen. As the surface atoms get converted to metal oxide, the section of the structural material keeps thinning and leads to possibility of failures. The oxidation rates can either be linear, parabolic or inverse logarithmic, depending on the substrate material and oxidation parameters.

2.2 *Sulfidation*

At high temperatures, sulfur in the environment reacts with the structural material (for e.g., Fe- or Ni-based alloys) to form low melting point eutectic sulfides such as FeS or NiS etc., which accelerates corrosion. The sulfur in the atmosphere may be in the form of polysulfides, aliphatic sulfides, hydrogen sulfide, disulfide, mercaptans, thiophenes, or sufur dioxide. In the presence of oxygen, the sulfidation attack becomes intensive with the formation of SO_2 and SO_3 gases.

2.3 *Hot Corrosion*

Presence of contaminants such as sulfur, sodium, potassium in fuels (coal or oil) results in intensified corrosion attack. The sulfur content in fuels along with alkali elements (Na, K etc.) in the presence of oxygen results in the formation of complex salts such as Na_2SO_4 or K_2SO_4 which gets deposited on the structural material. Temperature of 800–950°C is believed to be the melting temperature of these salts which accelerates the corrosion of structural material. This is called type I hot corrosion. The type II corrosion takes place at lower temperatures (670–750°C) wherein the sulfate deposits such as Na_2SO_4 or K_2SO_4 reacts with SO_3 on the surface of alloy to form pyrosulfates (e.g., $K_2S_2O_7$). These pyrosulfates have low melting point (670–750°C) and hence accelerate corrosion. Type I hot corrosion is observed in thermal power plants and gas turbine engines operating at high temperatures, while type II corrosion is observed mostly in thermal plants operating at lower temperatures.[1,2]

2.4 *Erosion and Erosion Corrosion*

2.4.1 Fly ash erosion

There are several kinds of erosion observed in thermal power plants and boilers such as fly ash erosion, soot blower erosion, falling slag erosion and coal particle erosion. Out of these, fly ash erosion is one of the most important erosions where the metal/alloy surface gets severely eroded. The main factor in the fly ash erosion is the quartz content in coal. The quartz particles passing through the flame remain relatively unchanged with a hardness of 1000–1200 kg/m^2. As a result, the tube material is directly removed due to abrasion, or indirectly damaged by removing the protective fireside scale which allows access of flue gases to the base metal.[3]

2.4.2 Erosion corrosion by liquid metals

The flow of liquid metals such as Pb–Bi through channels of metals/ alloys such as steel leads to penetration of Pb–Bi into steel through grain boundaries. Compared to the ordinary flow of fluids such as water, the liquid metals with high density like Pb–Bi exert 10 times higher dynamic pressure and shear stress on steel surface. Consequently, the mechanically weakened parts may be broken by the fluid mechanical forces.[4] Such a combined effect of erosion and corrosion at elevated temperature leads to intensified degradation of structural material.

2.5 *Carburization and Metal Dusting*

Carburization and metal dusting are key issues related to high temperature processing under carbon rich environments having high carbon potentials ($a_c > 1$).

2.5.1 Carburization

Carburization is a process of saturating the surface of a metal or alloy with carbon. Metals and alloys are susceptible to carburization when exposed to environment containing CO or CH$_4$ or other hydrocarbon gases at high temperatures (above 815°C). This is one of the major modes of high temperature corrosion in processing equipment of petrochemical industry. The carburized surfaces are brittle and tend to crack due to excessive internal stresses linked with carbon induced volumetric change and carbide precipitation.

2.5.2 Metal dusting

Metal dusting is a form of catastrophic carburization that occurs in steels and Fe, Ni, and Co-based alloys when exposed to process gas atmosphere of CO, H$_2$, and CO$_2$ with some hydrocarbons at 400–800°C temperature. At such high temperatures, CO and H$_2$ tend to dissociate on metal surfaces and form carbon. This carbon then diffuses into the substrate and pulls metal atoms out of their solid matrix, which leads to pitting and finally breakdown of materials in the form of dust.[5]

2.6 *Liquid Metal Corrosion*

When structural materials come into contact with liquid metals, they degrade by mechanisms of dissolution, alloying, mass transfer etc. For e.g., Ni dissolves in Pb or Pb-alloys owing to its solubility in Pb. This is why stainless steel (SS316) when exposed to Pb–Bi eutectic for long durations, results in formation of ferritic zones on the surface or sub-surface regions. Electrochemical techniques to measure corrosion rates are not applicable owing to the good electrical conductivity of liquid metals. Corrosion rates are thus measured by the extent of physical degradation. Dissolution, one of the modes of LM corrosion, is a temperature dependent phenomenon and hence the elements dissolved from the structural material in hot sections of nuclear reactor might precipitate out at the colder sections resulting in the choking of channels. The dissolution also leads to thinning of structural alloy sections leading to possibility of breakout. Use of inhibitors in liquid metals and coatings on structural materials are considered reliable approaches to reduce this type of corrosion.

3. Requirement of Coatings

The need for coating generates from the compatibility and functional requirements of the environmental & service conditions. However, the compatibility of such coatings with substrate is also critical and needs to be carefully addressed.

3.1 *Basis of Coating Consideration*

Selection of coating in the coating process is an important step for a design engineer and the basis of consideration is a decisive step. The basis of such selection is dependent on various factors which are as listed:

3.1.1 Metallurgical compatibility

The adhesion of coating with the substrate is very important and is linked with its metallurgical compatibility with the substrate. Diffusion bonded coatings such as aluminides are one such example of coatings with very good metallurgical compatibility with steels or superalloys. However, care has to be taken so that large amount of interdiffusion does not take place as it might lead to change in coating properties.

3.1.2 Mechanical compatibility

In order to maintain the protective feature of coating, its mechanical properties should match with the substrate. Factors affecting mechanical compatibility are coefficient of thermal expansion (CTE), ductility, cohesion, weight, surface roughness etc. For example, CTE is the main reason for oxide coatings delaminate or spall from the interface during high temperature exposure.

3.1.3 Coating process compatibility

The coating process should not induce properties undesired for end applications and hence it has to be compatible for targeted application. For example, an application demanding high dimensional tolerance cannot be subjected to a coating process conducted at high temperature which might induce distortions. The coating process should also not modify the microstructure of substrate material.

3.1.4 Component coatability

The type of coating process decides the technological feasibility of a coating. The coating process has to be selected depending on the geometry of the sample and the type of coating required. For e.g., a thermal spray coating cannot be chosen for jobs with intricate shapes or for coating inside surface of pipes with small diameter.

3.1.5 Environmental compatibility

The compatibility of coating with environment is the principal functional requirement without which the coating loses its purpose.

3.1.6 Available coating database and performance data

Database of information on coating properties and performance information is very important while selection of coating and coating processes. For example, growth rates of alumina thin films at different temperatures are well documented and can be used for projecting suitability of coating at selected operating temperatures.

3.1.7 Coating standardization

The selection of coating also depends on the standardization of the available coating process. Any variation in the process parameter may lead to change in coating structure and properties which might further lead to issues during performance. A standardized process parameter provides reproducible and reliable coating properties. Performance of coatings generated by a standardized process can be easily projected.

3.2 *Types of Coating Processes*

The process of generating a coating on substrate plays a vital role on the compatibility as indicated in Section 3.1. As a result, selection of the appropriate coating process is an important step. Depending

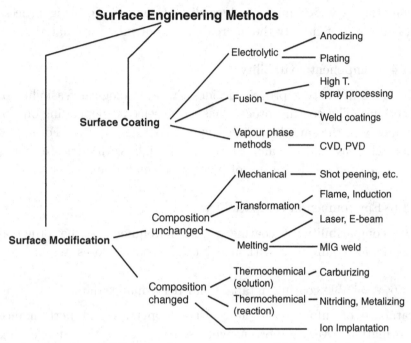

Fig. 1. Overview of different surface engineering methods, types of coatings and processes.

on the application requirement, the coating may be an overlay coating, or may be a thermochemically diffused coating. Different surface engineering methods and type of coatings are illustrated in Fig. 1.

Most of the coatings required for high temperature are oxides or intermetallics or phases of materials stable at such high temperatures. Out of the above referred coating and surface modification techniques, the most widely used processes for generating coatings for high temperature applications are thermal spray, laser surface alloying, diffusion coating processes and vapor deposition techniques.

The thermal spray process is a generic name of processes in which ceramic, metallic, or cermet powders in the form of wire, powder or rod are fed into a spray gun with which they are heated close to or above their melting points. The resultant molten droplets are sprayed through a gas stream onto a substrate to be coated. Thermal spray processing is a line of sight processing and

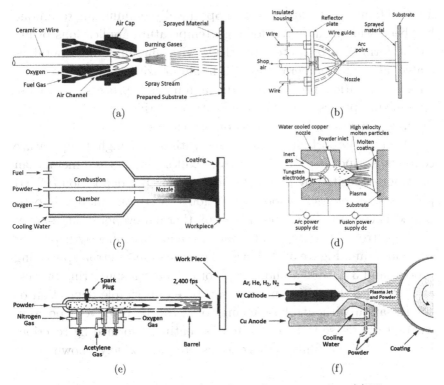

Fig. 2.   Schematic diagrams of different thermal spray processes (a) Flame spray (powder or wire type). (b) Wire arc spray. (c) HVOF spray. (d) Transferred arc spray. (e) D-gun. (f) Plasma spray.[6]

coating thickness is generated using multiple passes of the coating instrument. Thermal spray processes (see Fig. 2) include flame spray, wire spray, detonation gun (D-gun), plasma spray, high velocity oxy-fuel (HVOF), and cold spray techniques.

Major advantage of thermal spray processing is that almost all the materials that do not decompose on melting can be spray coated and that too without significant heating of the substrates to be coated. Additionally, ability to strip worn out surface and re-coat the substrates using thermal spray is a value addition. However, line of sight processing is a serious drawback involving complex shapes and non-accessible areas to be coated.

Diffusion coating processes such as aluminizing, chromizing, siliconizing etc., have been employed since decades as coating processes for high temperature applications. Specifically, aluminizing technique has been widely employed for high temperature applications owing to the formation of protective Al_2O_3 with aluminide intermetallics stable at high temperatures. Such aluminizing can be done by pack cementation process, hot dip aluminizing process, gas phase aluminizing process etc. The applications of this process have been discussed in subsequent sections of this chapter.

Another important process for synthesis of high temperature coatings is vapor deposition process. Oxides, cermets and metals can be vapor deposited by physical or chemical vapor deposition (CVD) processes. In physical vapor deposition (PVD), the material to be coated is converted to vapor phase by thermal evaporation, e-beam or magnetron sputtering under vacuum and subsequently deposited on a substrate. Figure 3(a) shows schematic of magnetron sputtering process and 3.3 (b) is a photograph of magnetron sputtering process at IPR. Such deposition process can be either simple deposition or condensation without phase change or it can be reactive deposition wherein the phase to be deposited as coating is formed by reaction. For e.g., alumina films, TiN coatings etc., are widely known to be

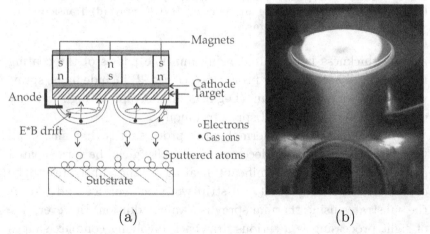

Fig. 3. (a) Schematic of PVD process by magnetron sputtering. (b) Glow discharge Ar plasma formed on target during magnetron sputtering [Photo courtesy: Institute for Plasma Research (IPR)].

deposited by reactive PVD process. In CVD, the source material to be coated on substrate is in gaseous precursor form which dissociates under high temperature and leads to deposition of the coating material.

4. Coatings for Thermal Power and Petrochemical Plants

Increasing the operating temperature capabilities of different steels for advanced power generation applications has been of prime importance due to potential gains in energy efficiency and contaminant decreases in emissions. However, such alloys are prone to oxidation at such high temperatures, and specifically the presence of water vapor or steam accelerates such oxidation process. Additionally, most of the thermal power plants are operated using non-renewable fuels such as coal or oil. This coal or oil fuel includes impurities such as sulfur, sodium and/or potassium, chlorine, vanadium etc. Hence hot corrosion and oxidation are important corrosion issues related to thermal power plants.

Coating requirement: The impurities along with sulfur and oxygen in atmosphere lead to formation of low melting point eutectic compounds on the surface of substrate which accelerates its corrosion and oxidation rates. A coating that is stable at high operating temperatures, resistant to attack of such low melting point eutectics, resistant to oxidation and which prevents migration of corrosive species into the bulk is desired. Additionally, such coatings should preferably self-heal after spalling. Applications relating to erosion–corrosion as discussed further in this section, demand erosion resistance properties in combination with the above referred high temperature corrosion resistance and self-healing properties.

Cr_2O_3 and Al_2O_3 are the most common corrosion resistant coatings meeting the above requirements. While Cr_2O_3 films have very good resistance to hot corrosion, their use beyond 850°C is not preferred as it transforms to volatile CrO_2 phase above 900°C thus rendering the substrate unprotected. Use of high Cr alloys may prove costly and hence diffusion coating processes such as chromizing on cheaper substrates are promising solutions. In thermal power plants, AISI 304 alloys may be the material of choice but the amount of Cr (18%) may be insufficient to form a protective and regenerating oxide on exposure to hot corrosion environment. Alloys with higher Cr content might prove to be uneconomical. In such cases, diffusion coating processes such as chromizing or chromizing-aluminizing become attractive choices.[7] Chromized steels (AISI 304, AISI 410) have been found to be protective for erosion corrosion resistance applications yielding lower metal loss than bare metals.[8] Chromized coatings have also been reported[9] to exhibit excellent resistance to molten salt (95% Na_2SO_4 + 5% NaCl) corrosion at 877°C as compared to bare AISI 410 and AISI 316. Such coatings have vital role to play in molten salt driven heat extraction systems in different types of power plants.

For high temperature applications, many applications demand aluminide coatings which yield self-healing alumina. In waste heat boilers utilizing biomass or municipal solid waste (MSW), the high temperature corrosion is one of the critical issues governing the energy efficiency. Such boilers are more prone to severe corrosion than other types of energy systems owing to deposits of alkali and heavy metals viz. Pb, Na, K, Cd, and Zn.[10] These elements originate from fuel and react with the environment and substrate to form low melting point eutectics, leading to hot corrosion. Alumina forming aluminide coatings, and chromia forming chromized coatings have been found effective in minimizing high temperature corrosion. Phongphiphat *et al.*,[10] compared the hot corrosion resistance of different alloys in a waste-to-energy boiler and observed that the order of hot corrosion resistance was as follows: aluminide coating Ni-based alloy 59 > aluminide coated SS310 > Ni-based alloy 625 > Ni-based alloy 59 > iron-based alloy 556 > SS316. HVOF sprayed NiCr10Al coatings showed excellent corrosion resistance in biomass boiler conditions for two years of exposure.[11]

Another important issue is the erosion corrosion of boiler components due to fly ash. As briefly explained in Section 2.4 of this chapter, the fly ash generated in coal fired boilers impinges on the surface of the boiler component and removes material, leading to

Fly ash, 540⁰C

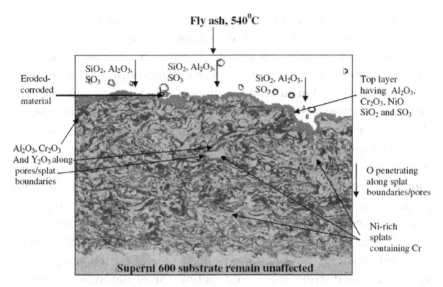

Fig. 4. Schematic diagram illustrating probable erosion corrosion of NiCrAlY coated superalloy exposed to boiler environment at 540°C for 1000 h.[1]

erosion. This fly ash also removes protective oxides and exposes bare substrate surface to the environment, thus enhancing corrosion. Mishra *et al.*[12] have reported the use of plasma sprayed NiCrAlY coatings for resistance to erosion corrosion in coal fired thermal power plants by exposure in actual boiler conditions at 540°C for 10 cycles of 100 h duration. Figure 4 gives an illustration of the fly ash erosion on NiCrAlY surface. SiO_2, Al_2O_3, and SO_3 from fly ash environment react on the surface, where initially oxides of Al, Cr, Ni etc. Further, the oxygen penetrates from pores/splat boundary forming oxides of Al, Cr, and Y which are protective in nature and block further penetration of oxygen. Such NiCrAlY coatings can also be deposited by other processes as described in this chapter.

In petrochemical plants and refineries, a combination of high temperature and hydrocarbon environment exists. The combination of hydrocarbon rich environment and high temperature leads to hydrogen embrittlement, carburization, decarburization, metal dusting etc., (see Section 2.5 of this chapter) depending on the application. A coating which resists diffusion or permeation of carbon or hydrogen from environment to the substrate and which is stable under such hydrocarbon environment at high temperature is required to mitigate such issues. Ceramic coatings stable at high temperature can be considered preferred candidates. In this context, Bayer[13] reported that alumina is more stable than chromia in high carbon activity environment at low oxygen partial pressures. Hence, aluminide or chromized-aluminized coatings are a preferred choice for carburization resistance and metal dusting in petrochemical plants. References 14, 15 reported the behavior of aluminide coatings on stainless steel component in methanol reforming plant in the temperature range of 565–1150°C. Such 120 μm thick aluminide coatings were reported to provide good resistance for 14 years in service with only occasional and minor pitting. Addition of Cr and Si in such aluminide coatings improves the performance of coated components against metal dusting and carburization. Si in aluminides reduces the diffusion of carbon from hydrocarbon environment to the base alloy. Aluminide (with Cr and Si) coated microalloy steel tubes were reported to withstand 27 months of service in ethylene furnace at 1149°C. Overlay coatings such as MCrAlY (M=Ni, Co), CeO_2 dispersed NiAl, glass coatings ($SiO_2+BaO+CaO+Al_2O_3$), cermets (Al/Al_2O_3, Cr/Cr_2O_3) etc., by different processes were also reported promising for petrochemical applications at high temperature.[5]

5. Coatings for Gas Turbines

Turbines need to operate at as high temperatures as possible to maximize their efficiency. A combination of high temperature and corrosive environment would not only degrade the mechanical properties of bulk material, but would also lead to catastrophic failure of turbine engine components due to corrosion.

Coating requirement: It would be desirous that the surface of turbine components such as blades and vanes has a 'coating' which confines the heat of the engine and acts as a thermal barrier (thermal conductivity preferably $< 3\,\text{W/mK}$). This would keep the bulk material at relatively lower temperature while operating at higher temperatures, and this enables them to take required mechanical load even at high temperatures. Such a coating would thus need to be stable at high operating temperatures ($> 1000°C$), sustain the stresses generated during heating and cooling, have low thermal conductivity, should not spall due to thermo-mechanical cycling, have matching coefficient of thermal expansion (CTE). The entire requirement of a thermal barrier coating system is thus divided into a top coat, a thermally grown oxide and a bond coat.

For top coat which imparts thermal insulation properties, complex oxides such as yttria stabilized zirconia (YSZ) have been found to be stable at such high temperatures and also provide very low thermal conductivity. Figure 5 indicates thermal conductivity of different insulator coatings as a function of time. As a result, YSZ coatings are preferred choice as thermal barrier coating (TBC) material.

TBCs are overlay type of coatings which are either spray deposited or vapor deposited on superalloy substrates. Such TBCs being oxides, they face major challenges at the coating-substrate

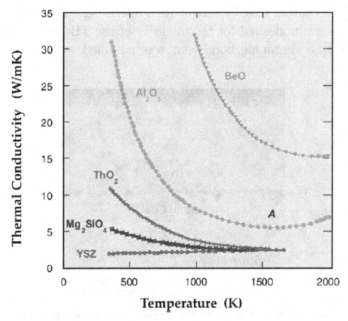

Fig. 5.   Thermal conductivity of different oxides as a function of temperature.[16]

interface. The sudden change in the compositional, physical and mechanical properties may lead to spallation of the coating. The difference in CTE between TBC (oxide) and substrate (metal/alloy) is one of the main criteria affecting such spallation. The role of a suitable interface coating between TBC and substrate which reduces the sharp CTE difference therefore becomes critical and inevitable. Such an interface coating should be a match between oxide and metals. High temperature stable intermetallics such as aluminides (NiAl, FeAl, TiAl etc.) are capable to form a protective alumina film on oxidation (also known as 'thermally grown oxide' or TGO). Such aluminides are thus considered as bond coat which have a graded Al diffusion profile on the substrate while a stable ceramic self-generated coating on the surface. This alumina (TGO) on the top of bond coat matches well with the YSZ thermal barrier coating. Since such aluminide coatings improve the bonding of TBC with substrate, they are known as '*bond coat*' (see Fig. 6). One of the important issues of aluminide coatings at high temperature is the formation of Kirkendall porosities formed due to internal diffusion of Al from coating to substrate at high operating temperature. Such porosities are undesired for the integrity of the TBC as a whole. As a result, the aluminide bond coat was modified with the addition

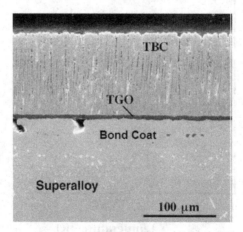

Fig. 6.   Cross-section image of YSZ thermal barrier coating deposited by e-beam physical vapor deposition on a superalloy.[16]

of Cr and Y elements. Addition of Cr facilitates the formation of stable α-Al_2O_3 film and Y acts as pegging agent and void sink thus preventing formation of Kirkendall porosity. MCrAlY type bond coats (where M = Ni, Fe, Co) have been established as reliable bond coats. Additionally, Pt-aluminide coatings or Hf-aluminide coatings are also reported to generate a very stable alumina film and that too with reduced growth rates.

The most critical component of the overall thermal barrier coating is the TGO layer at the interface between the bond coat and TBC which is also the weakest link. Most TBC failure studies are related to the delamination or failure of TGO from the bond coat, as visible from Fig. 7. On MCrAlY type bond coats, the TGO is alumina. It is important that out of all polymorphs of Al_2O_3, mestastable phases such as γ, θ, κ-Al_2O_3 be avoided and α-Al_2O_3 should be preferentially formed. The growth rate of metastable Al_2O_3 phases is higher compared to stable α-Al_2O_3 phase. As per Pilling–Bedworth ratio, beyond a critical oxide layer growth the TGO would tend to spall. Figure 8 indicates a typical TGO film thickness range beyond which spalling has been reported.[18]

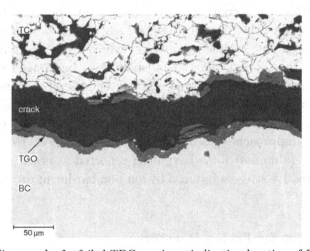

Fig. 7. Micrograph of a failed TBC specimen indicating location of failure close to top coat-bond coat interface or TGO–TBC interface. TC = Top coat or TBC; BC = Bond coat.[17]

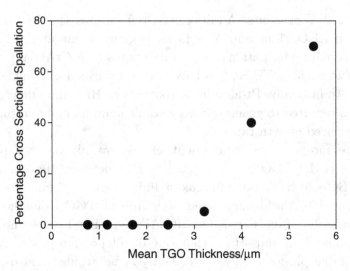

Fig. 8.   Percentage fraction of TGO spalled from a MCrAlY coating as a function of TGO film thickness.[18]

The TGO growth during service period should thus be restricted or very slow. α-Al_2O_3 which is stable and has slowest growth rate amongst all its polymorphs is hence desired. Another way to control the growth rate is to reduce oxygen availability at the TGO interface. The coating system should thus be blind to oxygen permeation. In other words, defects such as cracks and porosities in TBC promote oxygen permeation which leads to induction of stresses in the coating followed by spalling or buckling. Dense YSZ type coatings are thus preferred and can be generated using e-beam physical vapor deposition technique. Addition of additives such as Y, Pt, Hf etc., in the bond coat leads to a stable TGO (alumina) film. Novel approaches such as plasma assisted heat treatment to form TGO (alumina) films have been reported as crack free due to the compressive stresses induced by ion bombardment during plasma processing.[19]

6. Coatings for Fusion Reactors

Thermonuclear fusion reaction involves confinement of D-T plasma core at the center of a toroidal vessel. Temperature of over

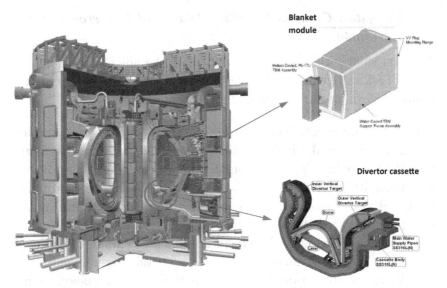

Fig. 9. Cutaway view of ITER model. Magnified schematic views of blanket module and divertor cassette provided with emphasis of their locations in ITER.[20]

5 million °C in the core of the reactor calls for confinement which can be done by gravitational, inertial or magnetic confinement. A joint effort is being made by a number of countries viz., China, India, European Union, Japan, Korea, Russia, and USA to setup an international thermonuclear experimental reactor (ITER) in Cadarache, France. An artist view of the reactor cross-section is given in Fig. 9.

At the center of the fusion reactor, the Deuterium (D)–Tritium (T) reaction would take place thereby generating Helium (He) (3.5 MeV) and neutrons (14.1 MeV). Material development challenges in a fusion reactor are mainly related to the development of materials which can face the harsh conditions posed by plasma and the highly energetic neutrons. While a lot of material development issues are involved in such fusion reactors, we are presently focusing on the surface, limited by the scope of this chapter. The assemblies of the fusion reactor such as divertor, first wall assembly and blanket module are critical components demanding surface engineering solutions in the form of coatings.

6.1 *Tungsten Coating for First Wall and Divertor Assembly*

The first solid surface of the reactor facing the hot plasma is known as the 'first wall', whereas divertor is a device or assembly that allows online removal of material and He ash from plasma.

Coating requirement: The first wall surface and divertor surface demands a material which can survive very high heat flux (ranging from 1–20 MW/m^2 depending on location of component). Additionally, the material of surface should not impact the plasma conditions. It should also demonstrate good bonding with the structural material, preferably a metallurgical bond.

These are also referred to as plasma facing components. Out of the choice of low Z materials (C, Be etc.) and high Z materials (W, W alloys etc.), tungsten has been the preferred choice, as far as ITER is concerned.[21] W eliminates the threat posed by tritiated radioactive dust generated by C-based first wall materials such as CFC or graphite. The tungsten is supposed to act as armor to the first wall and divertor assembly, thereby protecting subassemblies from thermal shocks and plasma effects. The typical heat loads on first wall is $< 1\,MW/m^2$ while that of divertor target plate is 10–$20\,MW/m.^2$ While the tungsten first wall will protect the subassemblies from such high heat loads, the structural material holding tungsten is required to be cooled. The first wall assembly of Indian test blanket modules (see Section 6.2) is cooled using high pressure He gas through cooling channels in the structural material i.e., reduced activation ferritic martensitic steel (RAFMS), while the divertor strike plate is fixed on a water cooled CuCrZr heat sink. Since Cu does not have adequate mechanical properties to survive irradiation environment, Cr and Zr additions make the CuCrZr a suitable candidate for divertor heat sink. W as a plasma facing material will act as an armor. W has to be either coated on first wall or has to be fixed as a tile on the first wall. In case of divertor, the plasma facing W material should have a very good metallurgical bond with the substrate (CuCrZr) and hence a coating of W is much more preferred.

Table 1:   Properties of W coating on CuCrZr alloy.[22]

Method	Thickness (μm)	Porosity (%)	Hardness (HV)	Oxygen content (wt.%)	Bond strength (MPa)	Thermal conductivity [W/(m K)]
LPPS	850	0.1	360	0.023	40.4	89.8
APS	1400	2.1	290	0.841	18.1	34.2
Cold spray	5	0.6	408	0	—	—
CVD	2	0	420	0.043	—	—
Electrodeposition	1127	0	423	0.018	50.5	93.4

W coatings by different processes such as low pressure plasma spray (LPPS), atmospheric plasma spray (APS), cold spray, PVD, CVD, and electrodeposition have been reported[22] for first wall applications. Table 1 gives an overview of W coating by different processes for first wall applications.[22] Reduction of porosity, improved adhesion with the substrate, low impurity content etc., are important factors for acceptability of coating and coating process. It can be seen from Table 2 that out of the thermal spray processes listed, the cold spray process indicates very low porosity levels and negligible oxygen content. The low oxygen content in cold sprayed W coatings is due to the low processing temperature ($\sim$500–600°C) wherein tungsten does not melt and hence chances of oxidation of W are very low.

Another promising technique is the electrodeposited W coating using molten salt electrolytes (e.g., $Na_2WO_4 + WO_3$) which has been reported by Refs. 23, 24 on Cu base alloys and V-4Cr-4Ti alloys respectively. The main benefit of this technique is the negligible porosity content and fairly thicker coatings in the range of few hundreds of microns, while the drawback is high temperature processing (1173 K) and long coating time (upto 100 h).

Since the CTE of W and CuCrZr is different, it would be desirable to have a graded coating. References 25, 26 have reported compositionally graded W-Cu coatings such that the W concentration decreases from surface to core while Cu content increases towards the substrate. This reduces the CTE mismatch between heat sink (CuCrZr) and armor material (W). Figure 10 indicates a

Fig. 10.   Cross section micrograph of a functionally graded W-Cu coating.[25]

cross-section image of FGM coating of W.[25] In order to improve the performance of W coatings, oxide dispersion strengthening (ODS) mechanism has been preferred and ODS–W alloys (with La_2O_3 or Y_2O_3) have been reported as PFC material.[27,28] La_2O_3 addition in W improves the grain boundary strength at ambient and elevated temperatures, and leads to improved thermal shock, creep resistance and machinability.[29] Coating processes that enable such ODS-W coatings with possibilities of functional and compositional grading would prove to be vital in this field.

6.2 *Insulator Coatings for Test Blanket Module*

The blanket module is the assembly of a fusion reactor which utilizes the energy of 14.1 MeV neutrons to generate tritium fuel which is critical for the economic feasibility of fusion reactor. Additionally, the blanket module also extracts heat out of the reactor which is utilized for generating electricity. As the name suggests, this system also 'blankets' the subassemblies and thereby protects them from radiation damage by slowing down energetic neutrons. The Indian test blanket module (TBM) considers breeder material in both solid (Lithium titanate ceramic pellets) and liquid (Eutectic Pb-17Li alloy) form. A schematic cross-section diagram of Indian TBM is given in Fig. 11.

The interaction of neutrons with Li in breeder material leads to generation of tritium, in both ceramic channels and liquid Pb-17Li flow channels. This radioactive tritium needs to be confined

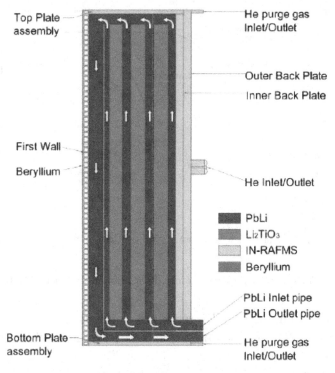

Fig. 11. Schematic diagram of Indian LLCB TBM concept.[30]

and should not diffuse into structural material (RAFMS). As shown in Fig. 12, coatings that can act as tritium permeation barrier have been found effective. The liquid breeder being a coolant also is an important part of the blanket, but also poses issues.

The liquid breeder use in blanket poses the following three main issues:

(i) Corrosion of structural material (RAFMS) by flowing Pb-17Li at 480°–550°C;
(ii) Permeation of tritium in structural material (RAFMS); and
(iii) Pressure drop due to magneto-hydrodynamic (MHD) effects.

Corrosion of steels by Pb-17Li has been one of the critical issues in liquid breeder blanket concepts. The liquid metal corrosion can take place in different forms viz., dissolution, alloying, intergranular

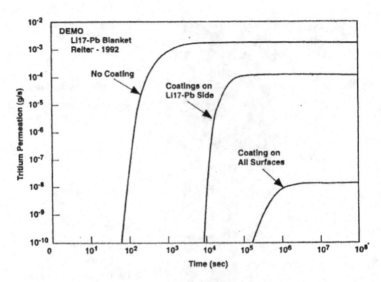

Fig. 12. Calculated T permeation into demo coolant from Pb-17Li blanket.[31]

penetration, impurity and interstitial reactions, mass transfer etc. Though dissolution is the most common form, more than one form of liquid corrosion takes place. Additionally, factors such as oxygen potential in liquid metal, temperature, liquid metal flow velocity, magneto-hydrodynamics etc., play a vital role in accelerating the degradation of steels.

Coating requirement: Based on the above information, the substrate needs to be isolated from liquid breeder by a coating which should be an electrical insulator (volume resistivity of coating $>10^4$ Ωm), should be resistant to tritium permeation (permeation reduction factor >100), should be compatible with Pb-17Li at 550°C under a velocity of 0.1–2 m/s for extended durations (>5000 hours) and should be self-healing in nature.

Insulator coatings such as Al_2O_3, Er_2O_3 etc. have been reported to be effective in resisting corrosion attack by Pb-17Li; reducing tritium permeation rates significantly; and in providing electrical insulation against MHD effects. However, a mere overlay of ceramic coating such as Al_2O_3 would create a large difference in CTE between the coating and substrate, which would possibly result in cracks.

Such cracks can become possible paths for accelerated permeation of tritium into RAFMS. As a result, a self-healing and graded coating such as aluminide (Fe-Al) has been the preferred choice.[32-34]

In recent work by the authors,[35] hot dip aluminizing followed by a plasma assisted heat treatment has been found to yield stable Al_2O_3 coating (3–4 μm) with FeAl diffusion coating of $\sim$70 μm thickness. The process of such coating involves aluminizing of 9Cr steels as first step, which is done by hot dip aluminizing or electrochemical Al coating. Subsequent to the formation of Al coating, the steel substrates are heat treated at 980°C for 30 minutes and at 750°C for 90 minutes. This heat treatment facilitates Al diffusion leading to preferred FeAl and Fe(Al) phase with a top Al_2O_3 layer. The bare 9Cr-1Mo steel (P91) and plasma treated aluminized P91 steel with α-Al_2O_3 coating were immersed in liquid Pb-17Li at 550°C for 1000 h under static conditions in ultra-high pure Argon atmosphere at 1.8–2 bar pressure. The positive Ar pressure would prevent possible ingress of oxygen from atmosphere into the chamber. Figure 13(a) indicates the extent of degradation of P91 steel substrates by Pb-17Li; and Fig. 13(b) indicates the protective nature of plasma grown Al_2O_3 film after 1000 h exposure to liquid Pb-17Li at 550°C under static conditions.

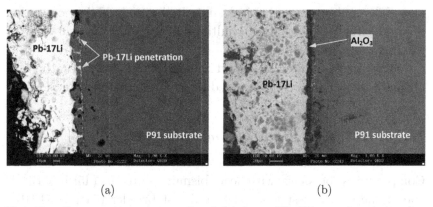

(a)   (b)

Fig. 13.   Cross-section micrograph of (a) bare 9Cr-1Mo steel sample after exposure to liquid Pb-17Li at 550°C for 1000 h; and (b) plasma grown alumina coating on 9Cr-1Mo steel after exposure to Pb-17Li for 1000 h at 550°C under static conditions.[36]

Aluminized coatings by conventional processes with a top Al_2O_3 coating have been reported to survive over 10,000 hours against flowing Pb-17Li and hence are important for blanket applications. However, further tests on the change in mechanical properties of aluminized RAFMS, irradiation properties of such aluminized coatings etc., need to be studied to accept this coating for application at site.

7. Coatings for Fission Reactors

Advanced nuclear power (fission) reactors and accelerator driven systems (ADSs) involve use of liquid lead (Pb) and lead-bismuth eutectic alloy as coolant and spallation target due to excellent physical, chemical and nuclear properties. However, these liquid Pb and Pb–Bi eutectic alloys are highly corrosive to most of the steels owing to high solubility of Fe, Cr, Ni in Pb/Pb–Bi. As referred in the previous section, the precipitation of the dissolved elements takes place at colder junctions of the heat exchanger. Liquid metal embrittlement is another important phenomena leading to degradation of mechanical properties of structural material.

Liquid sodium (Na) is another liquid metal used in the steam generator modules of fast breeder reactors. The steam generator material requirement is critical since it demands higher operating temperatures ($\sim$535°C) and simultaneously operates in contact with Na and water/steam.

Some issues related to liquid metal corrosion and coatings to mitigate them have been discussed briefly in subsequent sections.

7.1 *Corrosion of Steels by Liquid Lead Bismuth Eutectic (LBE)*

Compatibility of steels with lead bismuth eutectic (45Pb-55Bi) is one of the critical issues in ADS and fast breeder reactors (FBR). Protective oxide coatings with self-healing properties on steels are promising candidates in such environment with controlled oxygen content in flowing Pb–Bi eutectic. The attack by Pb-Bi on steels not

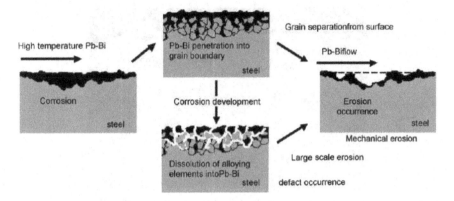

Fig. 14. Schematic of erosion mechanism of steel in flowing Pb–Bi.[4]

only involves corrosion by mechanisms stated in section 7, but also involves erosion of steel due to the high density of the flowing eutectic alloy.

Kondo *et al.*,[4] conducted an experiment liquid metal corrosion of different steels with flowing Pb–Bi at 550°C and at 0.5 L/min flow rate. As shown in Fig. 14, Kondo *et al.*, explained the mechanism of combined erosion and corrosion effects due to Pb–Bi flow. The Pb–Bi penetrates into the steel surface preferentially through grain boundaries. Due to this grain boundary penetration, the grains are weakened and taken away from the steel matrix due to hydrodynamic shear forces of Pb–Bi flow.

Another issue is the liquid metal embrittlement (LME) of structural material (steel). The outward dissolution of Fe from structural material leaves point defects which become sites for inward diffusion of Pb–Bi alloy. These sites contribute to the embrittlement of the steel structures.

The above referred corrosion attack by Pb–Bi can be mitigated by coatings that are self-healing and stable in Pb–Bi at operating temperatures. Aluminizing of steels has been considered a reliable coating process, which generates a stable Al_2O_3 layer over iron aluminide diffused layer on steel. Work reported by Ref. 37 on corrosion behavior of aluminized steels indicates how aluminide coatings can prove relevant for martensitic as well as austenitic steels.

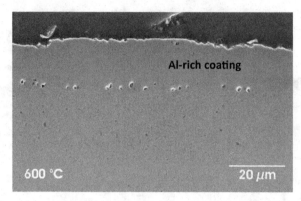

Fig. 15. Cross-section micrograph of SS 316L with Al surface alloyed, exposed to Pb–Bi at 600°C for 4000 h. No corrosion attack is visible.[38]

Compatibility of Al alloyed surface of SS316L with Pb–Bi has been reported in Ref.[38] with relevance to ADS, wherein the Al alloyed surface of SS316L has been observed to be immune to Pb–Bi at 600°C for 4000 hours while the unalloyed surface resulted in dissolution attack at 600°C. Figure 15 indicates a clean microstructure with undisturbed coating surface and core microstructure after exposure to Pb–Bi.

a. Fretting wear under liquid sodium environment

Fretting is a small amplitude oscillatory movement by the components of a system which are nominally at rest with respect to each other. In a prototype fast breeder reactor (PFBR), the steam generator (SG) module forms the boundary between liquid Na on reactor side and water/steam on plant side. The SG module involves steam carrying tubes made of P91 alloy (i.e., modified 9Cr-1Mo steel) and these tubes are immersed in liquid sodium. Since the steam inside the tubes is at very high pressures, such tubes experience flow induced vibrations (FIV), which are detrimental for reliable component service life. In order to damp such flow induced vibrations, clamps made of inconel 718 alloys are used to fix the tubes (see Fig. 16). However, the continuous vibrations result in fretting wear of the inconel alloy. Other components of the nuclear reactor

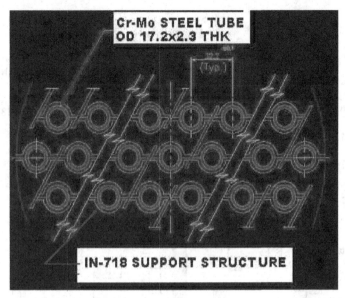

Fig. 16.   (a) Schematic arrangement of Cr–Mo steel tubes immersed in liquid Na and clamped using Inconel 718 alloy clamps.[39]

which experience fretting wear are fuel pins of reactor core, heat exchanger tubes against supporting components, holders, grids etc.

Initially, the SG design concept involved the use of steam carrying 9Cr-1Mo (P91) steel tubes supported by a $2\frac{1}{4}$ Cr – 1Mo steel grid plate immersed in liquid sodium. The flow induced vibrations in the steel tubes leads to galling and fretting wear of the grid plate and P91 steel tubes. Reduction of coefficient of friction (CoF) was required to reduce galling and fretting wear apart from a harder surface of grid plate.

Coating requirement: The coatings for the above said applications are expected to survive liquid Na environment at 625°C, thermal cycling from 200–625°C, neutron irradiation fluence of $6 \times 10^{22} \, n/cm^2$ or more. Additionally, the coating should have low CoF against $2\frac{1}{2}$ Cr – 1Mo steel in liquid Na.

Addressing the above issue, Lewis and Campbell[40] reported that use of aluminized inconel 718 alloy as clamps in SG modules which reduces the CoF, thereby reducing the fretting wear

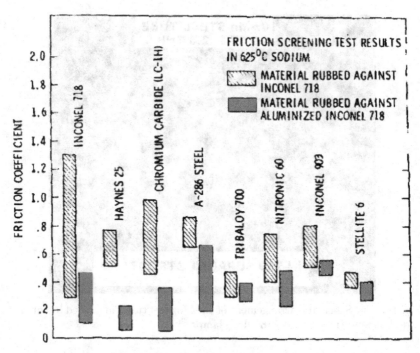

Fig. 17.　Effect of nickel aluminide diffusion coating on friction coefficient of alloy 718 rubbing against other materials in liquid sodium at 625°C.[41]

rates of 9Cr-1Mo steel tubes by a factor of 100 in comparison with unaluminized inconel 718 or $2\frac{1}{4}$ Cr – 1Mo steel counter faces. Johnson[41] illustrated a comparison of the CoF of different materials (see Fig. 17) in bare and aluminized condition and reported that aluminizing reduces the CoF. The aluminized In718 alloy which has a diffused NiAl coating is preoxidized in air at 700°C before exposure to Na. This leads to the formation of a top Al_2O_3 thin film. Subsequently, on exposing to liquid Na, the Al_2O_3 film reacts with Na to form $NaAlO_2$. This sodium aluminate has lower CoF and improves lubrication, thus resulting in reduced fretting wear.

NiAl can be formed by different processes as indicated in earlier sections of this chapter. Hot dip aluminizing is one technique which is capable to generate homogenous NiAl coatings on In718 alloys. Thermal spray coatings involve risk of porosities formed during spray process. Work done at the IPS involved hot dip aluminizing of In718

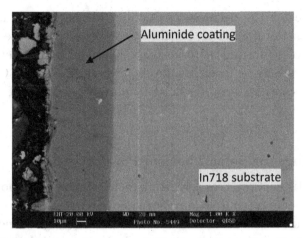

Fig. 18. Cross-section micrograph of plasma heat treated aluminized In718 samples at 950°C for five h. The coating shows homogenous nickel aluminide coating.

alloys leading to the formation of Al coating on In718 substrates. This Al coated substrates were then subjected to heat treatment at 950°C for five hours in vacuum furnace as well as in hollow cathode plasma environment. Figure 18 represents a cross-section of aluminized In718 substrate after heat treatment at 950°C for 5h under plasma processing. The surface to core elemental depth profiling using energy dispersive spectroscopy (EDS) on the coating visible in Fig. 18 confirmed the presence of NiAl phase as co-related with the Ni–Al phase diagram.[42] The solubility limit of elements such as Cr, Mo, Nb and Si in NiAl is very low and hence the NiAl aluminide coating may contain fine precipitates in the form of carbides or intermetallics.[43]

Fretting wear is also reported as an issue in accelerator driven systems (ADSs). Fretting in such ADSs is of concern as the resultant mechanical wear (fretting) can weaken or damage the corrosion barrier coating (e.g., alumina). This damage might reduce the service life of the components and an oxide scale breakoff may also result in choking of coolant channels.

In nuclear systems handling liquid metals, coatings are also important to resist liquid metal embrittlement (LME), which is responsible for the decrease in ductility of the core structure and also affects the ductile to brittle transition temperature (DBTT). Exposure of structural material (e.g., steels) to liquid metals in the form of breeders or coolants poses a threat of LME. Oxide rich protective coatings are beneficial in preventing LME of structural materials.

8. Coatings for Solar Thermal Power

The depleting oil reserves and rising fuel costs and greenhouse gases, renewable energy sources have become indispensable as alternate energy source. Activities related to wind energy, solar energy, tidal and hydropower energy have been looked upon with hope to reduce greenhouse gases and also meet the increasing demand of energy.

Solar energy can be tapped in various ways i.e., thermal and photovoltaics. The photovoltaic devices (solar cells) convert solar energy into electrical energy, while the solar thermal concentrators absorb the thermal energy of solar radiation and utilize it to produce steam which can be used for electricity generation as well as process heating applications. Since the present chapter is on high temperature coatings, we would limit our discussions to concentrated solar power (CSP) plants.

Concept of tapping solar thermal energy is as simple as visible from a magnifying lens used in sun to burn a matchstick or a piece of paper, i.e., concentrating the solar radiation at a focal point leads to an increase in heat flux. The means of concentration can be point focus or line focus. Parabolic dish or central receiver type solar thermal concentrators are point focus type, while parabolic troughs are line focus type. A typical parabolic trough is illustrated in Fig. 19.

A parabolic trough comprises a parabola shaped reflector (or mirror) which focuses the reflected solar radiation to the focal line; an absorber located along the axis of the focal point of parabola which absorbs all the solar energy; and a solar tracker to move the parabolic trough assembly in line with the sun to maximize efficiency. The mirror is required to be highly reflecting and its reflectivity should be close to 1 (ideally). Conventionally, the reflectors or mirrors

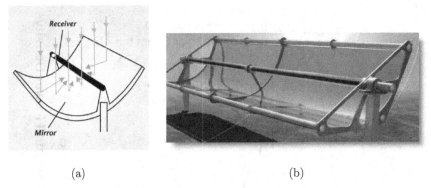

(a)   (b)

Fig. 19. (a) Schematic of a parabolic trough; (image source: _____) and (b) a picture of parabolic trough. (image source: _____)

are aluminized sheets with a protective transparent coating on the surface, while the absorber pipe (see black pipe at the axis of parabolic trough in Fig. 19) is a metal/alloy pipe (typically stainless steel or copper depending on type of concentrator) with a coating on the top to absorb the solar thermal energy reflected by mirror. The major efficiency loss is related to the absorber pipe. Depending on the heat transfer fluid temperature (T), parabolic troughs are of three types[44,45]:

(a) Low temperature ($T < 100°C$);
(b) Medium temperature ($100 < T < 400°C$);
(c) High temperature ($T > 400°C$).

In order to improve the efficiency, the absorber pipe should selectively absorb the solar radiation. The absorber coating can be classified into (i) non-selective coating and (ii) selective coating.[45] The non-selective coatings are those which have spectrally uniform optical properties, independent of waveform. Black paints, urethane paints with absorptivity (α) equal to 97–98% and emissivity (ε) ranging from 89–90% are non-selective coatings. Owing to high emissivity, the efficiency of such absorber coatings is low. As shown in Fig. 20, a solar spectrum from 0.25–2.5 μm wavelength needs to be absorbed. Beyond 2.5 μm, black body radiation starts which increases emissivity and is undesirable.

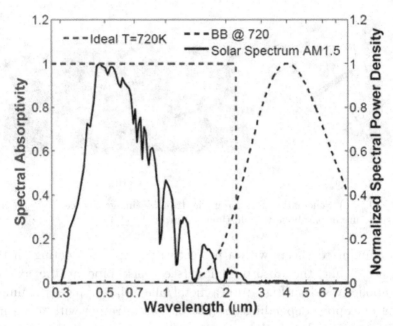

Fig. 20. Normalized spectral power density of black body (BB) at 720 K and solar spectrum. The spectral absorptivity is shown for an ideal absorber at 720 K.[46]

Coating requirement: The absorber coatings should possess improved absorptivity ($\alpha > 0.9$), low emissivity ($\varepsilon < 0.2$), long term stability of optical properties at operating high temperatures ($100°$–$500°$C) and environmental conditions, thermal stability of coating at high temperatures, and should be non-reactive to environment. Such absorber coatings, specifically used for mid-high temperature applications, are <500 nm thick. Most of the mid-high temperature absorber coatings are deposited by either physical vapor deposition (PVD) or chemical vapor deposition (CVD) techniques. At high temperatures, for a consistent long term performance, the absorber coatings have to be protected from reacting with the environment, from the erosion by the abrasive particles in environment. A schematic of the evacuated absorber tube cross-section for mid-to-high temperature applications is shown in Fig. 21.

Mid to high temperature absorber coatings have been of prime interest owing to the potential of efficient large scale solar power

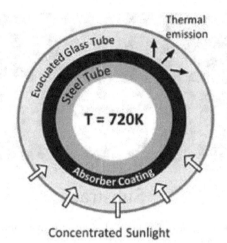

Fig. 21. Schematic diagram of evacuated absorber tube with solar selective coating.[46]

generation. Based on the optical and physical properties, a composite coating comprising metal and ceramic phases (also known as '*cermet*' coatings) have been found to be ideal for mid-high temperature solar thermal applications. While designing the absorber coatings, it is important to completely absorb the solar radiation, and to prevent heat loss by emission by the substrate held at high temperatures. An absorber coating thus has a sandwich structure as illustrated in Fig. 22.

As shown in Fig. 22, the top layer is an anti-reflection coating, typically a transparent oxide (ceramic) which absorbs the solar radiation without reflecting it back and thereby enhances absorption. The second layer is a cermet coating (metal-oxide composite structure) which absorbs the solar radiation, thus transferring the heat to the substrate. Subsequently, below the cermet and above the substrate is an IR reflecting coating meant to reduce substrate emittance. Absorber coatings with single cermet layers as indicated in Fig. 22 show a normal solar absorbance (α) of 0.8. In order to improve the absorbance ($\alpha > 0.9$), graded cermet coatings have been widely reported.[44,48] Zhang and Mills[48–50] suggested that the reflectance from the cermet absorber layer is reduced by gradually increasing the

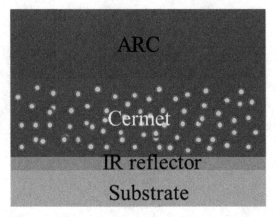

Fig. 22. Schematic diagram of a layered structure in cermet coating for solar thermal absorber coating.[47]

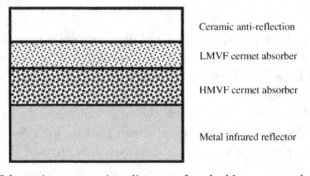

Fig. 23. Schematic cross-section diagram of a double cermet solar selective absorber coating.[48]

metal volume fraction of the cermet layer. A double cermet coating as illustrated by Zhang[48] is given in Fig. 23. Based on the layered structured explained in Fig. 23, Zhang and co-workers demonstrated Mo-Al$_2$O$_3$ cermet selective surfaces by vacuum co-evaporation which resulted in absorbance of 0.96 and near-normal emittance of 0.08 at 350°C when deposited on a Cu infrared reflector.

A variety of solar selective absorber coatings for low temperature, mid temperature and high temperature applications have been reported, some of which are listed in Table 2.

Table 2: List of absorber coatings by PVD for mid to high temperature applications.[44]

Type of coating	Substrate	Deposition process	Absorptivity (α)	Emissivity (ε)	Stability Temperature (°C)	Application
Black chrome	Cu	Vacuum evaporation	0.80	0.05 at 100°C	200°C in vacuum	Mid temperature
Cr_2O_3/Cr with Cr_2O_3	Ni plated SS	Sputtering	0.92	0.08 at 100°C	300°C in air	-do-
Metal carbides	Bulk Cu	Sputtering	0.76–0.80	0.02–0.03 at 100°C	400°C in air	-do-
Metal silicides	Bulk Cu	Sputtering	0.75–0.81	0.02–0.03 at 100°C	400°C in air	-do-
Al/AlN	Glass	Sputtering	0.93	0.04 at RT	400°C in air	-do-
TiN_xO_y	Cu	Sputtering	0.95	0.05 at RT	300°C in air	-do-
a-Si:C:H/Ti	Al	PVD/PECVD	0.75	—	250°C in air	-do-
$NiCrO_x$	SS	Sputtering	0.80	0.14 at 100°C	<200°C in air	-do-
Al_2O_3/Mo/Al_2O_3	Mo	Vacuum evaporation	0.85	0.11 at 500°C	400°C in air; 920°C in vacuum	High temperature
Pt-Al_2O_3 multilayers	Quartz	Evaporation	0.95	0.08–0.2 at 827°C	700°C in air (300h)	High temperature
W/W-Al_2O_3/Al_2O_3	SS	Sputtering	0.93	0.1 at 400°C 0.14 at 550°C	580°C in vacuum	High temperature
TiAlN/TiAlON/Si_3N_4	Cu	Sputtering	0.93–0.94	0.05–0.08 at 82°C	550°C in air; 600°C in vacuum	High temperature

While a number of such coatings have been commercially exploited, many of them are yet to be fully exploited on large scale. Techniques such as nano-patterning, plasma processing etc., would add value to the properties and performance of coatings.

9. Conclusion

Working of engineering systems at elevated temperature poses challenge not only from the strength perspective, but also from the reliability of extended service life of the engineering components. With the growing energy demand across the globe, discussions in this chapter were focused on coatings relevant to high temperature processing for different types of power plants viz., thermal power (fossil fired) including gas turbines, nuclear power (fusion and fission), and solar thermal power. Role of ceramics, cermets, intermetallics as high temperature coating materials in combination with high temperature structural alloys has become the focal point for the development of next generation power plants. Such high temperature coatings also play a vital role in other applications such as next generation automobile engines with reduced CO_2 emissions, futuristic space vehicles, aeronautical engines and vehicles, defense applications, hydrogen storage applications etc.

References

1. A.S. Khanna, Introduction to high temperature oxidation and corrosion (*ASM International*, Ohio, 2002).
2. G.Y. Lai, High temperature corrosion and materials applications (*ASM International*, Ohio, 2007).
3. D. Gandy, in: *Program on Technology Innovation: Stat of Knowledge Review of Nanostructured Coatings for Boiler Tube Applications, EPRI*, Palo Alto, CA, March 2007. 1014805.
4. M. Kondo *et al.*, *J. Nucl. Mater.* **343** (2005), pp. 349–359.
5. A. Aguero *et al.*, *Oxid. Met.* **76** (2011), pp. 23–42.
6. R.C. Tucker, Thermal Spray Coatings, in: ASM Handbook, *Vol. 5 Surface Engineering* (ASM International, Ohio, 1996), pp. 1448–1455.
7. F.J. Pérez *et al.*, *Surf. Coat. Tech.* **184** (2004), pp. 47–54.
8. Y. Liu and A. Levy, *Surf. Coat. Tech.* **43/44** (1990), pp. 836–847.

9. C.-T. Liu and J.-D. Wu, *Surf. Coat. Tech.* **43/44** (1990), pp. 493–499.
10. A. Phongphiphat *et al.*, *Corrosion Science* **52** (2010), pp. 3861–3874.
11. M. Osaka *et al.*, *Fuel Process. Technol.* **125** (2014), pp. 236–245.
12. S.P. Mishra, K. Chandra and S. Prakash, *Surf. Coat. Tech.* **216** (2013), pp. 23–34.
13. G. Bayer, *Corrosion 2001*, Nace International, paper no. 01387 (2001).
14. K.L. Baumert and J.J. Hoffmann, *Corrosion 1997*, NACE International, paper no. 493 (1997).
15. B. Ganser, K.A. Wynns and A. Kurlekar, *Materials and Corrosion* **50** (1999), p. 700.
16. D.R. Clarke and S.R. Phillpot, *Materials Today* **8** (2005) pp. 22–29.
17. R. Eriksson *et al.*, *Surf. Coat. Technol.* **236** (2013), pp. 230–238.
18. H.E. Evans, *Surf.Coat. Technol.* **206** (2011), 1512–1521.
19. N.I. Jamnapara *et al.*, *Surface Engineering* **30** (2014), pp. 467–474, doi: 10.1179/1743294414Y.0000000267.
20. A.K. Suri, N. Krishnamurthy and I.S. Batra, *Journal of Physics: Conference Series* **208** (2010) 012001.
21. M. Merola *et al.*, *Fus. Eng. Des.* **89** (2014), pp. 890–895.
22. Q.-Z. Yan *et al.*, *J. Nucl. Mater.* **442** (2013), pp. 190–197.
23. F. Jiang *et al.*, *Fus. Eng. Des.* **89** (2014), pp. 83–87.
24. Y.H. Liu *et al.*, *Fus. Eng. Des.* **87** (2012), pp. 1861–1865.
25. J. Song *et al.*, *J. Nucl. Mater.* **442** (2013), pp. 208–213.
26. C. Ge *et al.*, J. Nucl. Mater. 283–287 (2000), p. 1116.
27. M. Rieth *et al.*, *J. Nucl. Mater.* **432** (2013), pp. 482–500.
28. Q.-Z. Yan *et al.*, *J. Nucl. Mater.* **442** (2014), pp. S190–S197. Retrieved from http://dx.doi.org/ 10.1016/j.jnucmat.2012.11.046.
29. I. Smid *et al.*, *J. Nucl. Mater.* **258–263** (1998), pp. 160–172.
30. E. Rajendra Kumar, T. Jayakumar and A.K. Suri, *Fus. Eng. Des.* **87** (2012), pp. 461–465.
31. G.W. Hollenberg *et al.*, *Fus. Eng. Des.* **28** (1995), pp. 190–208.
32. G. Benamati *et al.*, *J. Nucl. Mater.* **271–272** (1999), pp. 391–395.
33. J. Konys, W. Krauss and N. Holstein, *Fus. Eng. Des.* **85** (2010), pp. 2141–2145.
34. W. Krauss *et al.*, *J. Nucl. Mater.* **417** (2011), pp. 1233–1236.
35. N.I. Jamnapara, PhD Thesis, Department of Metallurgical Engineering and Materials Science, Indian Institute of Technology Bombay, Mumbai, 2013.
36. N.I. Jamnapara *et al.*, *J. Nucl. Mater.* **464** (2015), pp. 73–79.
37. P. Deloffre, F. Balbaud-Célérier and A. Terlain, *J. Nucl. Mater.* **335** (2004), pp. 180–184.
38. G. Muller *et al.*, *J. Nucl. Mater.* **335** (2004), pp. 163–168.

39. G. Srinivasan, V. Shankar and A.K. Bhaduri, *Workshop on Steels and Fabrication Technologies 2008 (WS&FT'08)*, IPR, Gandhinagar, India, 2008.

40. M.W.J. Lewis and C.S. Campbell, Aluminized surfaces for prevention of galling and fretting wear in fast reactor steam generators, *3rd Int. Conf. Liquid Metal Engineering and Technology*, Oxford, U.K., 1984, p. 9590.

41. R.N. Johnson, *Thin Solid Films* **118** (1984), pp. 31–47.

42. G.K. Dey, *Sadhana-Acad. P. Eng. S* **28** (2003), pp. 247–262.

43. C.-J. Wang and S.-M. Chen, *Surf. Coat. Technol.* **201** (2006), pp. 3862–3866.

44. N. Selvakumar and H.C. Barshilia, *Sol. Energ. Mat. Sol. C* **98** (2012), pp. 1–23.

45. A. Abbas, *Renew. Energy* **19** (2000), pp. 145–154.

46. N. Sergeant and P. Peumans, *High-performance absorbers for solar thermal applications*, SPIE Newsroom, (2009) 1–3.

47. F. Cao *et al.*, *Ren. Energy Environ. Sci.* **7** (2014), pp. 1615–1627.

48. Q.-C. Zhang, *Sol. Energ. Mat. Sol. C* **62** (2000), pp. 63–74.

49. Q.-C. Zhang and D.R. Mills, *J. Appl. Phys.* **72** (1992), p. 3013.

50. Q.-C. Zhang and D.R. Mills, *Sol. Energ. Mat. Sol. C* **27** (1992), p. 273.

Chapter 8

Advanced Analytical Tools to Understand High Temperature Materials Degradation — Ion Beam Characterization of Aerospace Materials

Barbara Shollock* and David McPhall[†]

*University of Warwick, United Kingdom
[†]Imperial College London, United Kingdom

1. Introduction

Although expanding rapidly to include carbon-based composites and ceramic matrix composites, traditionally, aerospace materials consist of aluminum alloys for airframe applications and titanium alloys and nickel superalloys for engine components. As service requirements have become increasingly demanding, the boundaries between the alloys and their usage have become less distinct; for example, structural components may now be manufactured from titanium alloys. Regardless of their final use, the effects of processing on the microstructure and chemistry of these alloys and their long term stability in operating conditions must be understood. Electron microscopy has been a key tool in elucidating these aspects, but more recently, ion beam techniques, such as focused ion beam (FIB) techniques have been found to provide complementary information at a range of length scales. FIB techniques provide a route to examine the near-surface microstructure and with proper preparation, produce information with few, if any, artefacts. When a secondary ion mass detector (SIMS) is attached, the FIB SIMS becomes a powerful tool for chemical analysis at ppm levels.

2. Sputtering and Ion–Solid Interactions

The fundamental interactions of the primary ion beam and the sample are critical for all aspects of ion microscopy: milling, imaging, material deposition and FIB–SIMS. Consider Fig. 1, which shows the bombardment of a solid surface with a primary ion beam to cause sputtering. Sputtering is a physical phenomenon, whereby surface erosion takes place on an atomic scale. Initially, the primary ion impinges on the surface, resulting in a series of elastic collisions, in which momentum transfer takes place from the incident ions to the atoms of the target material. This causes atomic displacements within a collision cascade region that will result in the ejection of an atom as a sputtered particle, provided it has enough kinetic energy to overcome the surface binding energy. The statistical nature of sputtering indicates that the number of sputtered particles will increase with the number of collisions taking place closest to the surface. Note that not all collisions remain elastic and secondary electrons will be ejected, which are important for imaging.

The ejected particles can form positive and negative ions or simply remain as neutrals. The effectiveness of the primary ion beam to sputter charged particles is known as the ionization yield Y^+. Different studies have shown that the positive ion yield increases with ionization potential, whilst the negative ion yield increases with the

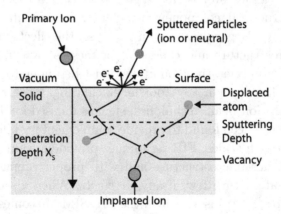

Fig. 1.   Schematic of the sputtering process and the collision cascade in the sub-surface of the target material.

electron affinity.[2,3] Furthermore, this parameter will depend on the matrix, the ion beam and residual elements in the vacuum such as oxygen.

Figure 1 shows another important parameter, the projected range X_S, defined by the perpendicular distance travelled by an implanted ion in a solid. This parameter is important because it reflects the magnitude of the stopping power dE/dx offered by the penetrated media; X_S is inversely proportional to dE/dx. The chance of the ejection of atoms increases if more collisions take place closer to the surface. In other words, less sputtering takes place if the ion penetrates deeper into the sample. Denser materials generally possess greater stopping power and accordingly demonstrate higher sputter yields.[4]

In most FIB instruments, primary ion beam production employs a Ga$^+$ liquid metal ion gun (LMIG), but gas sources, such as the Hyperion^{TM} are slowly emerging as alternative sources. Figure 2 presents a schematic diagram of a FIB SIMS instrument.

By far, the most utilized primary ion source, the Ga$^+$ LMIG, easily achieves beam currents of 5 nm, suitable for milling and imaging. The variety of beam currents achievable (1 pA–20 nA) allows users to conduct SIMS in the form of mass spectra, depth profiles and elemental maps. An electron multiplier and gas inlet needles are essential for the production of high quality images of FIB milled cross-sections. In short, the Pt gas injection source can be

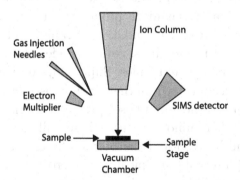

Fig. 2.   Schematic illustrating the basic setup of the FEI FIB200-TEM SIMS workstation.

used to lay a thin strip of Pt on the surface to protect the surface from unwanted sputter damage. The electron multiplier collects the secondary electron and positive ions to form detailed micrographs of the area of interest. Note that the ability to conduct milling, SIMS and microscopy *in situ* without the removal of the specimen from the vacuum between steps gives this setup a significant advantage over other SIMS instruments.

3. FIB Secondary Electron and Secondary Ion Imaging

In the scanning electron microscope, images generated by secondary electron (SE) images are typically used to examine the topography of the sample, while back-scattered electrons (BSE) are used to generate images containing information on composition and crystallographic orientation. With the advent of new in-lens detectors, these images can be obtained using lower accelerating voltages, eliminating the need for conductive coatings and increasing the resolution, but the interpretation of images can still be challenging. The use of complementary images produced using a primary ion beam can provide further information, particularly in samples with both electrically conducting and non-conducting phases, such as metallic alloys with an oxide scale. Nickel superalloys are used in the hottest parts of the gas turbine in the most challenging environments. Superalloy processing from the liquid state provides a key route to controlling final mechanical properties. Polycrystalline alloys, generally produced from consolidation of rapidly solidified powders, are used in the turbine discs, while single crystal castings are used for turbine blades and vanes. Superalloys derive their primary strengthening mechanisms from the γ/γ' microstructure. Of course, γ and γ' are often not the only phases present, and the polycrystalline alloys contain other minor phases. These will include the carbide phases (MC and $M_{23}C_6$) and the boride phases (MB$_2$ and M_3B_2).[5] Figures 3(a)–3(c) reveal the typical γ/γ' microstructure using various characterization techniques with increasing spatial resolving power of a polycrystalline superalloy produced by powder metallurgy and

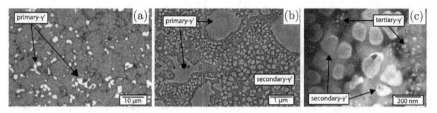

Fig. 3. Microstructure of a powder metallurgy nickel superalloy using different characterization techniques (a) light microscope, (b) SEM-SE microscope and (c) BF TEM microscope.

subjected to a heat treatment to optimize the microstructure. These micrographs illustrate the trimodal γ' distribution in the alloy. Figure 3(a) shows a typical light micrograph following etching with Kalling's solution (etches γ'). At this magnification, it is possible to identify the primary-γ' precipitates that reside at the grain boundaries of the γ matrix. Figure 3(b) shows the microstructure as viewed using SEM in the secondary electron mode after electrolytic etching in $10\%HP_3O$-$90\%H_2O$ (removes γ), revealing the coherent secondary-γ' particles in the γ matrix. Figure 3(c) shows a bright field (BF) transmission electron microscope (TEM) image, which identifies the even smaller tertiary-γ' precipitates distributed in the γ channel between the secondary-γ' precipitates.

The demand for increased engine efficiencies, whether to reduce costs or emissions, pushes the nickel superalloys to their mechanical and thermal limits. During the design of a gas turbine, much emphasis surrounds the *turbine entry temperature* (TET); the temperature at which the hot gases enter the turbine section. To maximize efficiency, and therefore, to reduce costs and greenhouse gases per unit of output,[5] it is desirable to increase the TET. To meet the demand of extended service life and lower operating cost, it is necessary to understand the behavior of these materials in conditions that match and exceed those experienced during normal in-service operating conditions. The development of an understanding of the oxidation and oxygen transport mechanisms in nickel superalloys will inevitably provide a basis for component lifing models for both polycrystalline discs and single crystal blades. Furthermore, an understanding of the degradation mechanisms of current alloys should only assist in the development of future alloys.

The oxidation resistance of turbine disc materials has generally received less interest over the years due to their relatively lower operating temperatures compared to the turbine blades; however, with increasing TET's for improved efficiencies, the effects of oxidation on disc alloys have become an increasingly important issue. Studies of the alloy surfaces exposed to high temperatures using SEM provide evidence of the oxide morphology and distribution; however, elucidating changes to the near surface provide a greater challenge. Conventional metallographic cross-sections require careful preparation to preserve the surface and near-surface microstructure.

High-resolution images of the surface can be obtained using FIB, using beam currents of typically <100 pA. For imaging purposes, the most important species are the secondary electrons and secondary ions. A typical FIB is fitted with an electron multiplier detector with the bias applied to this device determining whether positive or negative species are collected. As both images are "ion induced", from here on, the two imaging modes will be referred to as an ion induced secondary electron image (IISEI) or a total positive ion image (TPII). The attractiveness of this as an imaging technique stems from the different contrast mechanisms offered. Before discussing these mechanisms, consider the two cross-sections in Fig. 4, prepared by FIB milling of a specimen of the same powder metallurgy alloy presented in Fig. 3 after exposure at high temperature in an oxidizing environment. The specific details of the oxidation behavior are not discussed here; however, the two micrographs elucidate the two main contrast mechanisms: (i) channeling contrast and (ii) chemical contrast (sometimes called material contrast).

Channeling contrast results from ion channeling and is most prominent in the IISEI mode as shown Fig. 4(a). Channeling is the process whereby ions penetrate further along low index crystallographic directions compared to non-channeling directions in close packed materials such as metals.[9] This increases the range of the ions and shifts the collision cascade deeper into the sample. It follows that the secondary electron yield decreases when the ions channel

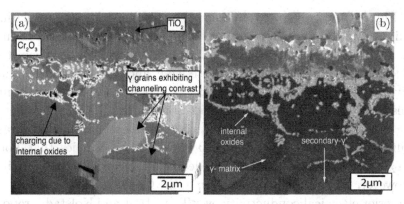

Fig. 4. Comparison of contrast mechanisms in FIB micrscopy using oxidized RR1000: (a) IIISEI and (b) TPII in an oxidized polycrystalline nickel superalloy.

further into the specimen because the electrons have less chance of escape.[9] Therefore, grains with low index directions aligned parallel to the incident beam appear dark. The Cr_2O_3 and TiO_2 layers formed during oxidation do not exhibit channeling contrast, even though they are polycrystalline, for the following reasons. First, both oxides do not possess close-packed structures, normally a prerequisite for channeling. Second, the oxides have much lower densities so that the collision cascade, and therefore electron generation, takes place deeper in the sample. Finally, the oxide becomes amorphous as it rapidly loses oxygen[10] and no channeling takes place. Channeling contrast can manifest in the TPII mode (see Phaneuf[11]); however, the number of electrons generated can be a factor of 10–1000 larger than that of ions, making it a less dominant contrast mechanism in TPII micrographs.

The second contrast mechanism observed relates to the chemical contrast, visible mostly in the TPII micrograph (Fig. 4(b)). Chemical contrast generally arises because sputter yields (i.e., S) depend on the target chemistry with denser phases (i.e., large dE/dx) generally appearing brighter. This reasoning should be treated with caution because the contrast is also related to the ionization yield. Take, for example, the effect of oxygen seen in Fig. 4(b), where yield enhancement leads to bright Al_2O_3 grain boundary oxides. Equivalent regions in Fig. 4(a) portray a very different effect for the oxide; the insulating oxides have a dark appearance. The positive ion bombardments of an insulator result in charge accumulation so that any emitted electrons will have no chance of escape and go undetected. This gives them a void-like appearance, which has led to them being mistaken for voids in previous oxidation studies.[12] Charging effects do not manifest in the external oxides so that chemical contrast differentiates the Cr_2O_3 and TiO_2 layers. This implies that both Cr_2O_3 and TiO_2 exhibit some degree of electrical conductivity. However, the brighter contrast seen in the Cr_2O_3 layer suggests it has superior electrical conductivity over TiO_2. Holt and Kofstad[13] report that Cr_2O_3 is, in fact, an electrical conductor, in-line with the current observations. Pure TiO_2 is nominally an insulator; however, it may develop non-stoichiometry (TiO_{2-x}), particularly in the presence of aliovalent cations, allowing it demonstrate n-type semi-conductivity. This could explain the conductive properties portrayed in the IISEI images.

Figure 4(b) readily shows the discernment of the γ'-precipitates amongst the γ matrix; the γ'-phase exhibits a greater total secondary ion yield relative to the γ-phase. To investigate the difference in contrast further, theoretical predictions of the sputter yield and the phase density have been made using transport of Ions in matter (TRIM) and JMatPro, respectively. Note that the JMatPro software also determines the γ/γ' phase partitioning using the base alloy composition. Table 1 gives the densities and sputter yields of the two phases, showing that the γ phase has a density greater than that of the γ'. Although the difference is very small, one would expect the γ

Table 1: Theoretical predictions of density and sputter yield for the γ and γ' phases.

Phase	Density ($g\,cm^{-3}$)	Total Sputter Yield (atom ion^{-1})*
γ	8.20	6.307
γ'	8.03	6.776

*Sputter yields assumed for 10^5 incident ions for a 30 keV Ga^+ at $0°$ incidence to the surface.

to have greater contrast over the γ'. However, theoretical predictions imply that the γ'-phase will have a higher ion yield.

The sputter yield alone does not explain this observation and analysis should also account for the ionization yields of each of the elements. Ionization yields for each element by means of Ga^+ are currently unavailable so the data provided by Storms *et al.*[3] for 13.5 keV O^- bombardment of the elements in their pure form have been used instead. Based on the SIMS equation for a species m:

$$I_s = I_P \rho_m S_m Y_m^+, \qquad (1)$$

where I_s is the secondary ion current, I_P is the primary ion current, S_m is the sputter yield, Y_m^+ is the ionization yield, ρ_m is the concentration of species m in the target and T is the characteristic transmission of the system. The final three terms in Eq. (1) can be considered as the useful ion yield. Ga^+ LMIG sources normally have a useful ion yield of the order 10^{-6}. Using this equation as applied to the phases present in the nickel superalloy, then the total number of ion for each element can be estimated from the product $S \times Y^+$. The sum of all elements in either phase gives the total ion yield (i.e., $\Sigma(S \times Y^+)$). Based on this, the total yield for the γ is approximately 1.82 times larger than that of the γ'. This result does not reflect the observations in Fig. 4(b). The enhanced secondary ion yield from the γ' must arise from some other mechanism. The enhancement may relate to residual oxygen in the FIB's vacuum chamber. Most FIB systems do not achieve vacuum pressures better than 1×10^{-6} mbar, and the possibility of oxygen arriving at the surface during analysis exists. As the ion beam rasters across each data point, the residual oxygen settles on the surface to increase secondary ion yields. Previous discussions described how Al, in particular, undergoes oxygen yield enhancement. Therefore, given the superior levels of Al in the γ'-precipitates compared to the γ-matrix (12.73 compared to 0.59), it appears that the brighter contrast arises from this mechanism.

The final contrast mechanisms concerns the cross-section of an oxidized sample presented in Fig. 5. Channeling contrasts persists both intergranularly and within the grains themselves. Previous

B. Shollock & D. McPhall

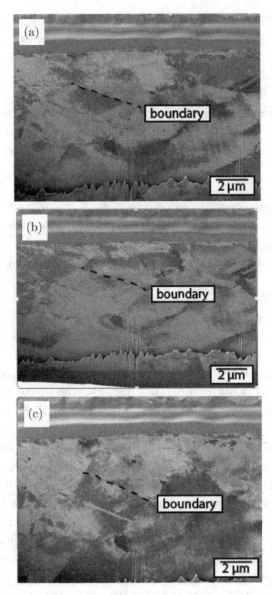

Fig. 5.   IISEI micrographs of shot-peened FGRR1000 at imaged at various tilt angles illustrating the effects of deformation on the channeling contrast: (a) 0°, (b) 7.5° and (c) −7.5°.

investigations[14] using EBSD have shown that it is possible for subgrains of different orientations to form through deformation. In the same manner that strong channeling exists between grains of large misorientation, small orientation differences, such as subgrains, are attributed to the more subtle changes in ion channeling and therefore, the contrast in the deformed regions.

For thin surface layers ($<2\,\mu$m), it can be difficult to distinguish the features in the layer in cross sectional images. In these cases, an alternative method of cross sectioning to increase the projection of the oxidation-induced damage in vertical scale allows for clearer images and a larger surface area for subsequent materials characterization. This can be achieved through shallow-angle beveling of the specimens using FIB milling. Figure 6 illustrates this schematically and demonstrates how milling at angle θ to the plane on the surface produces a cross-section that exaggerates the true thickness x_{true} of the oxidation damage.

The projected thickness as viewed in the FIB x_{viewed}, assuming a zero stage tilt, is related to the milling angle θ by:

$$x_{\text{true}} = x_{\text{viewed}} \tan(\theta). \qquad (2)$$

From this simple relationship, it is easy to calibrate the vertical scale for any cross-section milled at all angles. The tangential nature of Eq. (2) implies that very large projections of the oxidation damage when applying shallow milling angles. The implementation of this

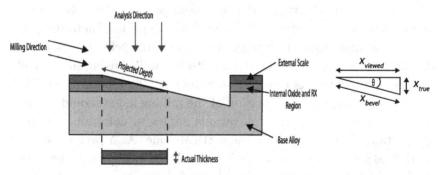

Fig. 6. Schematic of the FIB beveling method used in the preparation of cross-sections for various characterization techniques.

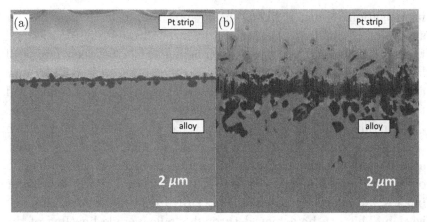

Fig. 7.   Cross-sectional microstructure of the thin oxide layer on an oxidized single crystal superalloy showing (a) milled at 90° angle and (b) beveled at 5° angle acquired back-scattered mode in the SEM.

methodology was particularly important for FIB–SIMS mapping methods this methodology increases the equivalent number of data points in a single pixel by the same factor. Figure 7 illustrates the power of this approach. In this figure, the thin oxide layer formed on a single crystal superalloy is shown in BSE images of trenches milled at 90° to the sample surface and with a 5° bevel. The beveling stretches out the oxide layer, stretching it out by about 10 times the cross-section thickness of layer.

The actual method to prepare cross-sections involves first depositing a thin Pt strip with dimensions of approximately $20\,\mu\mathrm{m} \times 3\,\mu\mathrm{m} \times 1\,\mu\mathrm{m}$ on the surface. This provides both protection to the oxide from unwanted sputter damage and a suitable reference point for thickness measurements. The trench itself is milled using successively smaller beam currents typically starting from 20 nA down to 0.5 nA for final polishing. Determination of the milling angle should be done by successive milling trials at various angles to determine the best projection of the features in the vertical scale. As a prerequisite for FIB–SIMS mapping, consideration should be taken as to the depth and width of the trench to avoid artefacts in the SIMS data due to loss of signal due to interaction of the sputtered ions with the

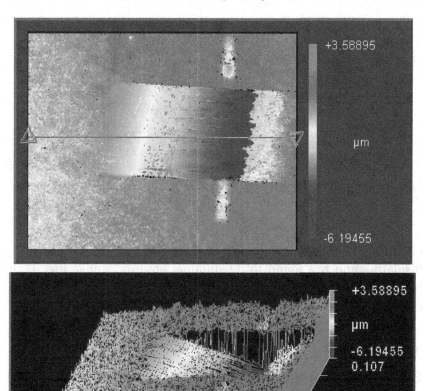

Fig. 8.   White light interferometer images illustrating the nature of the bevel milled into the surface of the oxidized single crystal nickel superalloy in 2–(top) and 3–(bottom) dimensions.

crater wall. Figure 8 shows the final geometry of the milled section presented in Fig. 7.

Figure 9 shows a further example demonstrating the effectiveness of shallow milling through the oxide on an oxidized single crystal nickel superalloy along with the TPII and IISI images contrast. In

Fig. 9. Comparison of contrst mechanisms in FIB micrscopy using oxidized RR1000 (a) IIISEI and (b) TPII in an oxidized single crystal nickel superalloy.

common with Fig. 4, the darkly imaging layer in Fig. 9(a) is highly insulating and was found to be alumina; however, unlike Fig. 4, the polished matrix region at the top of the section shows no channeling contrast because of its single crystal nature.

4. FIB–SIMS

FIB–SIMS is a versatile materials characterization technique, which combines the sensitivity of SIMS with the excellent spatial resolution, imaging and milling capabilities of FIB methods. On its own, SIMS is a surface analysis technique that analyzes the chemical characteristics of the surface through mass spectrometry of the sputtered ions.[15] Using FIB–SIMS, users have the ability to analyze the region of interest (ROI) by mass spectrometry, depth profiling and elemental mapping.

The attractiveness of SIMS stems from its parts per million (ppm) to part per billion (ppb) sensitivity (i.e., the minimum concentration that generates one count per data point)[16] for relatively large area analysis; however, this can drop considerably for much smaller areas, especially for LMIG sources.[17] This describes an underlying issue faced by SIMS users: the inverse relationship portrayed by sensitivity and analytical volume. Referring back to Eq. 1, the final three terms can be considered as the useful ion yield. Ga^+ LMIG sources normally have a useful ion yield on the order 10^{-6}, whereas reactive sources such as oxygen and caesium possess useful ion yields of around 10^{-3}. It follows that the sensitivity also decreases by three orders of magnitude when using FIB–SIMS.[9] At higher dopant concentrations or analytical volumes, this may not pose as much of an issue; however, for FIB-SIMS analyses requiring <100 nm beam diameters for high-resolution mapping, this quickly becomes problematic.

SIMS analyses allow elemental analysis at ppm levels while FIB milling allows preparation of challenging samples. An example of such a sample is a metal matrix composite — a relatively soft metal containing hard ceramic particles. These materials are important in aerospace applications where strength-to-weight ratios to be optimized to deliver the performance needed. In these composites, the relative differences in hardness and electrochemical response makes imaging the microstructure using conventional techniques difficult due to specimen preparation artefacts produced during traditional metallographic preparation routes.[11] Figure 10 presents a FIB polished surface secondary ion image of solid state processed Al2124 alloy with boron carbide reinforcement. The boron carbide reinforcements can be seen as light regions on the order of 1 micron in the gray matrix. In addition, isolated bright spots are visible.

The SIMS spectra in Fig. 10(c) reveal differences between the analyzes of the bright spots and matrix area. There is a clear difference in the peak intensity at mass 24 amu with small difference at 25 amu and 26 amu, all corresponding the Mg isotopes ^{24}Mg (abundance 79%), ^{25}Mg (10%) and ^{26}Mg (11%). Based on calculations of the alloy composition, the area fraction of bright particles and thermodynamic data, the composition of the bright spots was determined to be $MgAl_2O_4$.[12]

FIB–SIMS has been applied to high temperature exposed samples of both single and polycrystalline nickel superalloys to elucidate the oxidation behavior. The focus of the FIB–SIMS study was not composition but instead involved investigation of the oxidation mechanisms using isotopic tracers.[13] Although FIB–SIMS suffers from sensitivity issues, as a mass spectrometry technique, it has the ability to readily differentiate between isotopes such as ^{16}O and ^{18}O, making

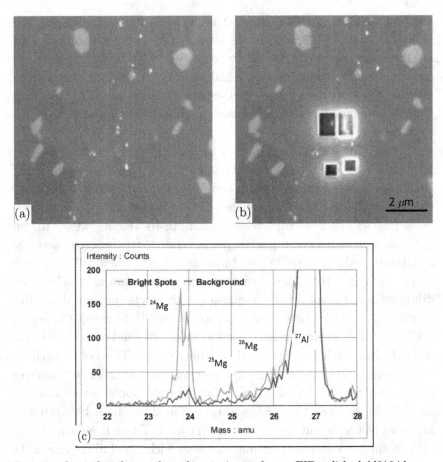

Fig. 10. Ion induced secondary electron image from a FIB polished Al2124 boron carbide composite (a) as-polished and (b) showing milled craters resulting from analysis of bright spots and adjacent matrix material (c) representative SIMS spectra from the bright spots and from the adjacent matrix.

it a useful technique in isotopic tracer experiments. Considering Eq. (1), for two oxygen isotopes used for isotopic oxidation, only the relative concentration affects the secondary ion signal, I_s. With the knowledge that negative oxygen ions generally have high yields, then FIB–SIMS is the ideal candidate technique to analyze the relative distributions of $^{16}O^-$ and $^{18}O^-$ in isotopic oxidation specimens.

References

1. R.C. Reed, *The Superalloys: Fundamentals and Applications*, (Cambridge University Press, Cambridge University Press, New York, 2006).

2. A. Benninghoven, F. Rudenauer and H.W. Werner, *Secondary ion mass spectrometry: Basic concepts, instrumental aspects, applications and trends*, Mass Spectrometry; Basic Concepts, Instrumental Aspects, Applications and Trends, (Wiley, New York, 1987).

3. H.A. Storms, Evaluation of a cesium positive ion source for secondary ion mass spectrometry, *Anal. Chem.* **49** (1977), p. 2003.

4. J.F. Ziegler, M.D. Ziegler and J.P. Biersack, SRIM–The stopping and range of ions in matter, *Nuclear Instruments and Methods in Physics Research Section B: Beam Interactions with Materials and Atoms.* **268** (2010), pp. 1818–1823.

5. D.J. Young, *High Temperature Oxidation and Corrosion of Metals*, Elsevier Corrosion Series, Amsterdam, (2008).

6. T. Gabb, J. Telesman, P. Kantzos, J. Smith and P. Browning, *Effects of high temperature exposures on fatigue life of disk superalloys* (2004), pp. 269–274.

7. T. Gabb, J. Telesman, P. Kantzos, J. Sweeney and P. Browning, *Effects of high temperature exposures on fatigue life of U720* (2002), pp. 261–269.

8. C.K. Sudbrack, S.L. Draper, T.T. Gorman, J. Telesman, T.P. Gabb and D.R. Hull, Oxidation and the Effects of High Temperature Exposures on Notched Fatigue Life of an Advanced Powder Metallurgy Disk Superalloy, *Superalloys* (2012), pp. 863–872.

9. L.A. Giannuzzi, B.I. Prenitzer and B.W. Kempshall, *Ion-solid interactions*, in: Introduction to focused ion beams, (Springer, 2005), pp. 13–52.

10. J. Telesman, T.P. Gabb, Y. Yamada, L.J. Ghosn, D. Hornbach and N. Jayaraman, Dwell notch low cylce fatigue behaviour of a powder metallurgy nickel disc alloy, *Superalloys* (2012), p. 853.

11. M. Phaneuf, Applications of focused ion beam microscopy to materials science specimens, *Micron.* **30** (1999), pp. 277–288.

12. J.R. Silk, R.J. Dashwood and R.J. Chater, FIB SIMS analysis of an aluminium alloy / SiC metal matrix composite, *Surface and Interface Analysis* **43** (2011), pp. 488–491.

13. A. Holt and P. Kofstad, Electrical conductivity of Cr_2O_3 doped with TiO_2, *Solid State Ionics* **117** (1999), pp. 21–25.

14. R. Hjelen, E. Orsund and E. Nes, On the origin of recrystallisation textures in aluminium, *Acta Metallurgica et Materialia.* **39** (1991), pp. 1377–1404.

15. D. McPhail, Applications of SIMS in Materials Science, *J. Materials* **41**(3) (2006), pp. 873–903.
16. D. McPhail, R. Chater and L. Li, Applications of focused ion beam SIMS in materials science, *Mikrochim. Acta.* **161** (2008), p. 387.
17. L. Li, Strategies for improving the sensitivity of FIB-SIMS, *Surf. Interface. Anal.* **43** (2011), p. 495.
18. R.D. Evans and J.D. Boyd, *Scripta Mater.* (2003), pp. 45–49.

Chapter 9

Role of Nanotechnology
in Combating High Temperature Corrosion

R.K. Singh Raman*, B.V. Mahesh[†] and Prabhakar Singh[†]

*Department of Mechanical and Aerospace Engineering,
Department of Chemical Engineering,
Monash University, Melbourne, Australia

[†]Centre of Clean Engineering, University of Connecticut, USA

Role of nanocrystalline and ultra-fine grained metals/alloys in improving the high temperature corrosion/oxidation characteristics is presented. The primary content of the chapter is an elaborate description of the authors' recent hypothesis and its experimental validation that a nanostructure can bring about dramatic improvements in the oxidation resistance of low chromium iron-chromium alloys at moderately high temperatures. A nanocrystalline Fe-10wt.%Cr alloy was found to undergo oxidation at a rate that was an order of magnitude lower than its microcrystalline counterpart. Importantly, the oxidation resistance of nanocrystalline Fe-10wt.%Cr alloy was comparable with that of the common corrosion-resistant microcrystalline stainless steels (that have 18–20 wt.% chromium). However, the chapter also provides a brief description of the difficulties in the processing of nanocrystalline iron-chromium ferritic alloys as well as the success in circumventing these difficulties. To set the scene on the oxidation resistance of nanocrystalline alloys, the chapter presents a summary of the recently reported roles of nanocrystalline structure in oxidation resistance of a few other chromia and alumina forming systems. The chapter also includes a brief description on the mechanism of oxidation resistance of alloys simultaneously having with both nano- and microcrystalline structure (i.e., bimodal structure).

1. Introduction

Single or multiphase polycrystalline solids with grain size typically less than 100 nm are known as nanocrystalline (nc) materials.[1,2] Because of their extremely fine grain size, a remarkably high volume of the nc materials is composed of interfaces (grain boundaries and triple points).[3,4] As a result, those properties that depend on the grain size and grain boundaries may be considerably different in the case of a nc metal/alloy as compared to its conventional microcrystalline (mc) counterpart. For example, nc metals and alloys exhibit increased mechanical strength, enhanced diffusivity and higher specific heat and electrical resistivity. Several such applications require the materials to demonstrate acceptable levels of resistance to environmental degradation. A proper understanding of environment-assisted degradation of nc metallic materials metals/alloys is particularly important since grain size and grain boundaries are known to influence corrosion processes. Oxidation/corrosion resistance of nc metals and alloys has received very limited research attention. However, nc metals and alloys have been reported to exhibit different oxidation/corrosion resistance to their mc counterparts.[5-7] It is also emphasized that besides the interest in investigating the role of nc structure in oxidation/corrosion, another aspect is the possibility of exploiting the enhanced grain boundary phenomenon (such as diffusion) for the purpose of developing oxidation/corrosion resistant alloys with considerably less alloying contents.[5]

Nanocrystalline materials have presented far superior alternatives to conventional alloys in a variety of commercial applications. For instance, the choice of materials for high temperature inductor applications for power generation, conversion, and conditioning is limited to those that possess sufficiently high Curie temperature and high magnetization at the operation temperatures. Nanocrystallization has been suggested as a solution for improvement of high temperature soft magnetic materials having low hysteretic loss and improved high frequency response.[8,9] In the clean energy industry, metallic interconnects for the intermediate temperature solid oxide fuel cells (IT-SOFCs) have been envisaged as a potential application for nc ferritic stainless steels.[10] Similarly, nc Cu-Ni-Cr alloys have been shown as a significant improvement over their mc counterparts for use in condenser tubes in marine environments.[11] Lightweight Ti-Al alloys find extensive application in aircraft and automobile industries. However, the poor oxidation/corrosion resistance of Ti-Al remains a technological challenge. Nanocrystalline coatings based on γ-TiAl, TiAlCrAg and AlCoFeCr which do not alter the bulk properties have been reported to improve the oxidation resistance of TiAl alloys.[12,13] Recently, Chookajorn *et al.*,[14] reported a solution to particularly vexing problem of nc materials — microstructural instability at elevated temperatures. Their thermodynamic calculations on a series of binary alloys determine the stability not only against grain coarsening but also phase separation. A new tool is now available for designing nc alloys that meet the high temperature requirement under service conditions. The W–Ti alloys are particularly attractive for use in turbine blades. As the research in the area of nc materials progresses, their use in several other high temperature industrial applications will become viable.

This chapter discusses the current state of research in high temperature oxidation of nc materials. Though ferrous systems (primarily iron and iron-chromium alloys) are the major materials covered in this chapter, the reported literature on the oxidation resistance of other systems has also been reviewed. This chapter also provides a brief description of the thermal stability and the challenges in synthesis of nc metals and alloys, and the attempts for circumventing these challenges.

2. Thermal Stability and Synthesis of Nanocrystalline Materials

Despite a few orders of magnitude difference in the grain size and dimensions of grain boundaries, the structures of nc and mc materials have been suggested to be similar.[15–20] However, the grain boundary

energy and degree of disorder in case of nc materials is significantly higher.[21-25] As a consequence, nc materials are exceedingly
susceptible to thermally assisted grain growth.[15,26] According to the
well-known Gibbs–Thomson equation, the thermodynamic driving
force for grain growth process in conventional polycrystals can be
expressed as,

$$\Delta\mu = \frac{2\gamma V_a}{d},$$

where V_a is the atomic volume, γ is the interfacial energy, and d is
the crystallite size. Due to the presence of high density of interfaces
and nanometer-sized grains in nc materials, it is expected that the
grain growth will be quite high. Contrary to such expectations, experimental observations have indicated that most nc materials exhibit
inherent grain size stabilities up to reasonably high temperatures
$(0.4–0.5\ T_m)$,[27-31] viz., Fe and Fe-based alloys resist grain growth up
to 400–600° C.[15,30,32-37]

The primary factor which influences the grain growth in conventional polycrystalline materials is the atomic diffusion at the
grain boundaries, and the kinetics of grain growth is governed by
two significant parameters, namely, activation energy (Q) and the
grain boundary exponent (m).[26,29] A number of investigations have
suggested two primary mechanisms for controlling grain growth:
(a) reduction of grain boundary mobility (kinetic stabilization), and
(b) reduction of the driving force for grain growth (thermodynamic
stabilization). The kinetic stabilization operates via various mechanisms such as Zener drag and pinning of the grain boundaries by
pores, inclusions, triple junctions or segregation of lower density
solutes at the grain boundaries.[26,27,29,30,38] The thermodynamic
stabilization can be accomplished by addition of a solute which lowers
the grain boundary energy upon segregation.[7,39-41]

Several techniques have been employed for producing nc solids
in powder and thin film forms. Inert gas condensation,[42-45] pulsed
plasma deposition,[28] and sputtering[43] have exclusively been used for
processing thin films or small amounts of nc materials. Electrodeposition[32,33,46] and severe plastic deformation[47-49] have been recognized

as the two relatively successful routes for processing nc materials in bulk quantities. Pulsed electrodeposition[46] has been employed, most notably, for synthesis of nc Ni–Fe and Ni–Co alloys.[50,51] However, synthesis by electrodeposition often requires use of additives (for biasing nucleation over growth) which remain in the material as impurities. Among the plastic deformation techniques, advanced ball milling is the most widely used technique for producing artefact-free nc powders. Groza[52] has reviewed various techniques employed for nc powder compaction, viz., high pressure/low temperature compaction, *in situ* consolidation,[53] hot compaction,[54] hot rolling,[55] explosive compaction,[56] high pressure torsion (HPT),[57,58] equi-channel angular pressing (ECAP),[59,60] equi-channel angular extrusion (ECAE)[61] etc. A few other techniques have been employed for consolidation into bulk nc alloys, such as shockwave sintering,[62] self-propagating high temperature synthesis (SHS),[63] arc-plasma sintering,[64] and spark plasma sintering (SPS).[65]

3. Oxidation Resistance of Nanocrystalline Metals/Alloys

3.1 *General Principles*

During oxidation, binary alloys with certain alloying elements (viz., Cr, Al and Si) can form a continuous layer of the chromia, alumina or silica, conferring substantial oxidation resistance. This has formed the basis of the development of common oxidation resistant alloys, such as stainless steels. Formation of a continuous layer of surface oxide is called external oxidation.[66] If, on the other hand, the inward flux of oxygen exceeds the outward flux of solute during oxidation process and the oxygen partial pressure (pO_2) at the oxide alloy interface is only sufficient to selectively oxidize one of the alloying elements, isolated oxide particles form within the sub-scale. This phenomenon is called internal oxidation.[66] For external oxidation and formation of a continuous layer of chromia, alumina or silica, a critical concentration of solute is required, which can be calculated by Wagner's treatment[66] for various systems/conditions. Critical amount of a solute for such a transition depends directly on its diffusivity in the alloy, besides other factors (viz., concentration of solute element, diffusivity in the oxide scale, temperature, etc.). The extremely fine grain size and the high volume fraction of grain boundaries of nc materials[15] can cause an extraordinary increase in diffusivity, and nc structures may have beneficial effects in the development of the protective oxide layer. For example, oxidation resistances of Fe-Al and Fe-B-Si alloys in the nc state are reported[67,68] to be superior than in their mc state. This behavior is attributed to Al and Si, the well-known protective oxide film formers, being the predominantly diffusing species respectively in the two alloys, and the nanostructure facilitating their diffusion and expedited formation of protective films (of Al/Si oxide).

3.2 Oxidation Resistance of Nanocrystalline Fe–Cr Alloys

In the temperature regime of 300–400°C, chromium diffusion in a nc Fe–Cr system is reported to be about four orders of magnitude greater than in a mc Fe–Cr alloy.[69] Given the extremely fine grain size and the resulting high diffusivity in nc Fe–Cr alloys, one would expect the Cr concentration required for internal-to-external transition to become substantially lower. Singh Raman et al.,[70] proposed the hypotheses: (a) the oxidation resistance of a nc Fe–Cr alloy should be considerably superior to a mc alloy of the same composition, and (b) it should be possible to attain remarkable oxidation resistance at considerably lower chromium contents of nc Fe–Cr alloys, as opposed to the considerably greater minimum chromium contents (13–15 wt.%) required for development and maintenance of the Cr_2O_3 layers in common stainless steels. In this context, it is noted that though the minimum chromium content for the development of Cr_2O_3 layer is 13% for ferritic alloys, and approximately 15% for austenitic alloys, common stainless steels contain at least 18 wt.% chromium, in order to provide sufficient chromium in the subscale for the purpose of self-healing in the event of disruption in the initial protective layer of Cr_2O_3.

Singh Raman *et al.*,[5,70] have recently reported the role of nc structure in remarkably improving oxidation resistance. In this context, it may be imperative first to have an overview of the role of grain size in the development of the protective layer of oxidation resistant oxide (which is generally the inner layer of the multilayered oxide scale). As suggested earlier, for a given combination of alloy-environment-temperature, where predominantly diffusing species can form a protective film and provide oxidation resistance, a decrease in grain size will facilitate protective film formation. Iron-chromium alloys (such as stainless steels) are the most commonly employed corrosion resistant mc materials. A common high-Cr alloy (such as 18Cr–8Ni stainless steel) forms during oxidation an inner layer of Fe/Ni oxide that eventually converts into a protective oxide, i.e., Cr_2O_3 when sufficient Cr diffuses from the alloy bulk to the oxide scale-alloy interface.[97] The kinetics of transition of Fe/Ni oxide into protective layer of Cr_2O_3 depends on the supply of chromium by diffusion in the alloy matrix, which is governed profoundly by grain size of the alloy. As clearly demonstrated in the literature,[71] a fine grained ($\sim$17 μm or less) 18Cr alloy easily developed a uniform layer of Cr_2O_3. For the same alloy with grain sizes greater than $\sim$40 μm, this protective layer of Cr_2O_3 was difficult to form during air-oxidation, as the inner layer of $(Fe, Cr)_3O_4$ continued to grow due to insufficient chromium supply.[72] Low-chromium Fe–Cr alloys fail to form a protective layer of Cr_2O_3.[72–75] Singh Raman *et al.*,[74,75] have investigated the role of grain size (15–60 μm) in oxidation resistance of such low-Cr alloys. Grain boundary diffusion in mc low-chromium alloys is never enough for the formation of a contiguous protective layer of Cr_2O_3. In fact, the alloy suffers predominant and extensive oxidation along grain boundaries, and a decrease in grain size rather increases the grain boundary internal oxidation.

The most common and simple testing for oxidation rate of metals and alloys at elevated temperatures is the determination of weight gain per unit surface area with time. In order to compare the influence of nc *vis-à-vis* mc structure on oxidation, nc and mc Fe–10%Cr alloy powders were produced by ball-milling, compacted

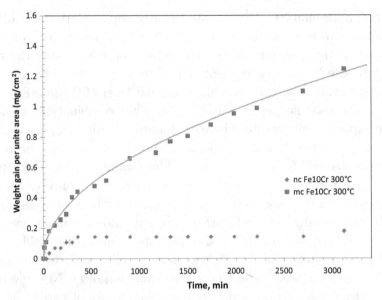

Fig. 1. Oxidation kinetics of nc and mc Fe-10Cr alloys oxidized at 300°C for 3120 min.[5,70]

into pellets and sintered.[5,76] The pellets of both nc and mc materials were oxidized at 300–400°C. Typical oxidation kinetics data (Fig. 1) show the mc alloy to be oxidizing at a considerably greater rate than the nc alloy of same composition. The striking features of the oxidation kinetics and oxide scale are:

- After 3120 min of oxidation at 300°C, weight gain of mc was found to be nearly an order of magnitude greater than that of nc alloy of same composition. This influence was more pronounced at 350°C.
- Both nc and mc alloys follow parabolic kinetics in the initial period (as was evidenced by the weight-gain^2 versus time plots). But, during subsequent oxidation, nc Fe–10Cr alloy showed a considerable departure from the parabolic behavior whereas mc alloy continued to follow parabolic kinetics.
- The considerable difference in the oxidation kinetics between nc and mc alloys after the initial period was also manifested in some stark difference in the color of the oxidized samples.

In this study, it has been possible to provide a concrete under-standing of the remarkable difference in the oxidation kinetics as well as systematic evolution of the oxide scale. Interested readers are encouraged to refer to an elaborate description cited in Ref. 5. However, a brief description of the oxidation kinetics and color evolution at 300°C is provided below.

In order to investigate the reason for the considerable difference in oxidation rates (Fig. 1) of the nc and mc states of the same alloy composition (Fe–10%Cr alloy), oxide scales were characterized. As described earlier, the oxidation resistance of Fe–Cr alloys is associated primarily to the chemical characteristics and the Cr content of the thin inner oxide scale. Depth profiles for Cr, Fe and O were generated using secondary ion mass spectroscopy (SIMS). Thin oxide films formed over nc and mc Fe-10 wt%Cr alloy during air-oxidation (30 min/300°C) were characterized by SIMS depth profiling. Typical depth profiles for Cr, O and Fe for the oxidized specimens are presented respectively in Figs. 2(a)–2(c). A comparison of the Fe, O and Cr profiles for the two specimens suggests the oxide film developed on mc Fe-10Cr alloy to be considerably thicker. Fe and O profiles of the two low-Cr alloys would suggest greater Fe and O contents in the case of the outer scale of nc Fe–10Cr alloy. The greater content of Fe-rich oxide in nc Fe–10Cr alloy is attributed to a greater grain boundary oxidation in the initial stages, owing to the far greater grain boundary area at the surface of a nanometric grain size material as compared to the mc alloy of the same composition.

The most notable finding of the SIMS profiles is that the Cr content of the inner layer of nc Fe-10Cr alloy is considerably higher (>4 times) than the Cr content in the inner layer of mc Fe-10Cr alloy. The remarkably higher Cr content of the inner oxide layer on nc Fe-10Cr alloy (as established in Fig. 2(a)), has been found to be comparable to the chromium content of the inner oxide layer that develops on a Fe–20Cr mc alloy[5] (which is well known to establish a protective layer of Cr_2O_3 at such temperatures). Thus, it was inferred that the nc Fe–10Cr alloys developed a protective oxide layer of Cr_2O_3, in spite of its considerably low chromium content.

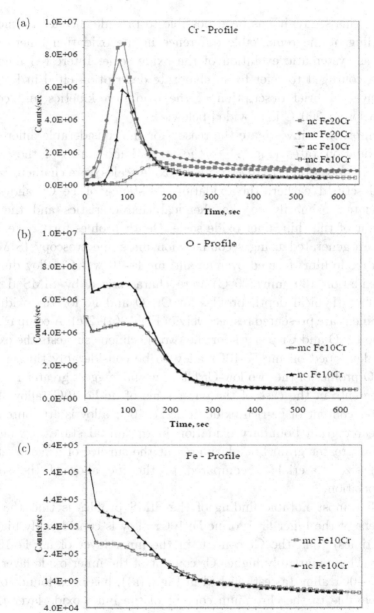

Fig. 2. SIMS depth profiles for Cr, O, and Fe in the oxide scales developed during oxidation of nc and mc Fe-10%Cr and Fe-20Cr alloys oxidized for 3120 min (a) 300°C, (b) 350°C, (c) 400°C.[5,70]

As the reported literature[72-75] would suggest, inner oxide layer of the mc alloys with low-Cr (<12%) would at best be a mixed spinel type Fe–Cr oxide. Such low chromium alloys would fail to develop a Cr_2O_3 layer.[72] It seems the remarkably greater oxidation rate of mc Fe–10Cr alloy (as compared to nc Fe–10Cr alloy) can be explained on the basis of the considerably higher Cr content of the inner oxide layer (Fig. 2) and possible development of the protective oxide layer of Cr_2O_3 over the latter. As a result of the considerable difference in chemical composition (chromium content) and the thickness of the oxide scales developed over the nc and mc Fe–10Cr alloys, the color of the pellets of the two materials oxidized for over 3000 min is distinctly different (Fig. 1 (insets)).

Comparison of the SIMS depths profiles for oxygen for nc and mc Fe–10Cr alloys (Fig. 2(b)) suggests the oxide scale developed over nc alloy to have: (a) less thickness (as a result of less effective protective scale, as discussed above), and (b) somewhat greater overall oxygen content (on the basis of the areas under the two profiles). However, a careful comparison of the oxygen profiles with Fe and Cr profiles (Figs. 2(a)–2(c)) would suggest that the location of the peak for O corresponds more to the peak for Fe in nc alloy, owing to the greater grain boundary oxidation of Fe in the early stages of oxidation of nc alloy. At the peak location for Cr, the oxygen content is considerably less. So the apparently greater oxygen content of the peak is largely associated with the greater grain boundary oxidation of Fe in the early stages. The eventual diffusion-assisted establishment of Cr_2O_3 layer ensures the oxidation rate of this alloy to be considerably lower. On the other hand, as the O, Fe and Cr profiles for mc alloy would suggest, the inner scale to the mixed oxide of Fe and Cr can provide considerably less protection than the inner of Cr_2O_3 that develops on the nc alloy, as described earlier.

For developing an understanding of how the considerably greater oxidation resistance of nc Fe10Cr alloy (in comparison with mc Fe-10Cr alloy) compares with the resistance of an alloy with much higher Cr content, samples of nc and mc Fe–20Cr alloys were also oxidized at 300, 350, and 400°C for durations up to 3120 min. The weight gains of these alloys at 300°C were too low to be detected by the gravimetric balance used for this study. However, it was possible to characterize the chemical composition of the thin oxide scales developed at the two temperatures. Oxidation kinetics of nc and mc Fe-20Cr alloys at 350°C[5] suggested only a little improvement in oxidation resistance due to nc structure. However, what is most relevant to note is that Cr contents of the inner oxide scale at the end of 3120 min-oxidation of mc Fe-20Cr alloy were respectively similar to those of the nc Fe-10Cr alloy at 350°C, suggesting the degree of oxidation resistance conferred due to nc structure at only 10% chromium to be similar to that of the alloy with twice as much chromium but mc structure.

Fe–Cr alloys with considerably low chromium content have demonstrated a remarkable oxidation resistance at 300–400°C, as a result of the nc structure.[5] The nc structure enhances diffusion and thus accounts for the improved oxidation resistance. In simplistic terms, one would expect this trend to continue with increasing temperature. However, the stability of the nc structure at higher temperatures may become an issue. Another factor will be balance between the increasing growth rate of non-protective external oxide and chromium supply. This aspect needs to be investigated in detail.

4. Potential High Performance Surface Coatings

nc metals and alloys in the bulk form exhibit poor ductility, especially under tensile conditions. This is, in part, due to the inability of nc grains to store dislocations. In contrast to the coarse-grained metals, the nc materials undergo an initial stage of rapid strain hardening over a very small plastic strain regime (~1–3%), beyond which there is very little or no strain hardening.[77-79] Consequently, this leads to localized deformation and the occurrence of necking quite close to the yield point and the net result is low tensile ductility.[80,81] Moreover, processing artefacts such as porosity, insufficient bonding between laminated layers and oxide/nitride contamination can also impair ductility. This will be a major limitation of nc materials in structural applications, despite their excellent oxidation resistance. To circumvent the issue of low ductility in nc materials, two approaches are suggested:

- Incorporation of micron-sized grains in the nc matrix to enhance their ductility (notably, the alloys with a bimodal grain size distribution possess remarkable oxidation resistance, in spite of having considerably less nc content, as discussed later).
- Since oxidation process is constrained to the surface and a few microns in the sub-surface, nc materials, with considerably less Cr content yet similar oxidation resistance as stainless steels could be used as inexpensive coatings for mitigation of oxidative degradation.

Several studies have shown that high temperature oxidation resistant coatings can be developed by the surface modification to a nc domain.[82,83] The selective oxidation of the coated layer to form an oxide phase which imparts the maximum oxidation resistance has been successfully exploited in various studies.[84–88] It has been established in the literature that nc coatings of superalloys developed by magnetron sputtering show excellent oxidation resistance in comparison to the mc alloy substrates of same composition as the coatings. For example, K38G (4%Al and 16% Cr),[85,86] KF17 (4%Al and 12.5%Cr),[84] and LDZ125 (5.2%Al and 9.5%Cr)[87,88] deposited on the substrate of the mc counterparts of these coatings form a mixture of Cr_2O_3 and Al_2O_3. The cross-sectional micrograph of the oxide films (Fig. 3) formed on a nc and as-cast K38G superalloy evidences that the selective oxidation of Al is preferred over the oxidation of Cr in the case of nc alloy, resulting in a unary layer of Al_2O_3.[89]

Improvement due to nc structure of coatings has been reported also by Liu *et al.*,[90–92] in the case of a Ni-20Cr-2Al superalloy. The coating with the smallest grain size (65 nm) showed a considerable improvement in oxidation resistance whereas the coating with an average grain size in the range of 280–350 nm had a resistance

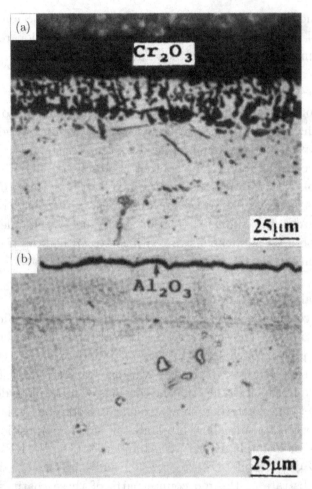

Fig. 3.   Cross-sections of specimens oxidized at 1000°C (a) cast K38G alloy for 200 h and (b) K38G nc coating for 500 h.[89]

similar to that of the uncoated bare alloy. A mechanism of selective oxidation of Al, to develop a well-established Al_2O_3 layer, despite the significantly low Al content (as compared to Cr) is proposed on the basis of the nanometric grain size facilitating the "third-element effect".[93] During the initial stages of oxidation of Ni-Al-Cr alloys, oxides of all the elements (Ni, Cr, and Al) are formed, largely in the ratio of their bulk concentrations. Due to the enhanced diffusion in nc matrix, a continuous thin layer of Cr_2O_3 (which is much less defective) is formed instead of a Ni–Cr spinel which is seen in the coarse-grained alloys. The Cr_2O_3 defect structure affects the diffusion/ion exchange through the oxide layer such that the oxygen partial pressure at the oxide-alloy interface becomes equal to the equilibrium partial pressure required for the exclusive formation of Al_2O_3. Thus, an Al_2O_3 sublayer forms and grows into a continuous layer.[94]

Improved oxidation resistance of the nc coatings as compared to the mc cast alloys of same composition can also been attributed to the spallation behavior.[83] The scale adhesion in case of nc alloy coating is significantly better as compared to coarse grained alloys. Due to the several-orders-of-magnitude difference in the grain size, the oxide scale is expected to be finer in nc alloys. The fine grained oxide layer is more capable of accommodating the growth stresses during isothermal oxidation as well as the thermal stresses during cooling by plastic deformation.[83] Furthermore, pegging of the oxide layer via "micropegs" (Fig. 4) anchors and enhances the mechanical bonding of oxide to substrate in the nc alloy.[83]

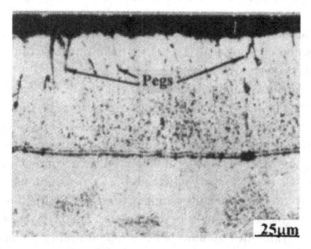

Fig. 4. Cross-section of $Ni_3(AlCr)$ nc coating after oxidation at 900°C for 200 h showing "micropegs".[83]

5. Oxidation Resistance of Bimodal Fe–Cr Alloys

Incorporation of micron-sized grains in a nc matrix (thus leading to a bimodal/multimodal grain size distribution) can be an effective way to improving the ductility of nc alloys.[95] This strategy for imparting ductility is motivated by imparting some ability for dislocation movement and strain hardening through the coarser grains, but inevitably lowers the strength of an otherwise fully nc material.

Recently, it has been demonstrated that bimodal grain size distribution can be exploited for improving the ductility of Fe–Cr alloys, without compromising the excellent oxidation resistance which is observed in completely nc structure of same composition.[96,97] The mechanism of ductility improvement and enhanced plastic deformation in bimodal structures is beyond the scope of this book. However, the oxidation resistance of bimodal Fe-alloys is discussed. In the study, Fe–Cr alloys with varying percentages of Cr and Zr and having three different kinds of grain size distribution (completely nc, bimodal and completely mc) have been compared. Figure 5 shows the oxidation kinetics of these alloys. It is evident from Fig. 5 that the oxidation kinetics of nc alloys is largely similar to that of the mc 20Cr alloy (in spite of their much lower Cr contents, only 7 and 10 wt.% Cr), and the Zr addition does not affect the oxidation rate. However, an interesting observation is that the bimodal alloys possess the level of oxidation resistance which is similar to that of the nc alloys of same composition (i.e., their oxidation resistance is much superior to the mc 10Cr alloy).

The elemental composition analysis of the oxide scale along the depth is shown in Fig. 6. The relative positions of the chromium peak are indicative of the oxide scale thickness whereas the height of this peak suggests the maximum chromium content at the oxide-alloy interface. Despite the lower bulk Cr content of bimodal alloys, the Cr enrichment (indicated by Cr-peak intensity) at the oxide–alloy interface is quite similar to that for the high Cr mc 20Cr and significantly superior to mc 10Cr alloy. Apparently, the nanometric grains in the alloys with bimodal grain size distribution facilitate sufficient Cr enrichment at the oxide-alloy interface for the formation of a protective layer of chromia (and hence good

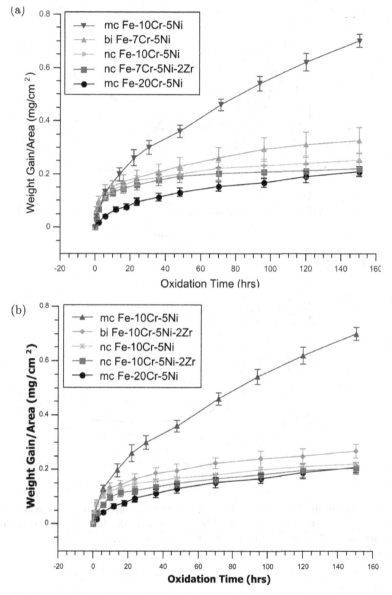

Fig. 5. Comparison of the weight gain during high temperature oxidation of various Fe–Cr–Ni–Zr alloys (a) nc and bimodal 7Cr alloys and (b) nc and bimodal 10Cr alloys, compared with mc alloys containing 10% and 20% Cr.[97]

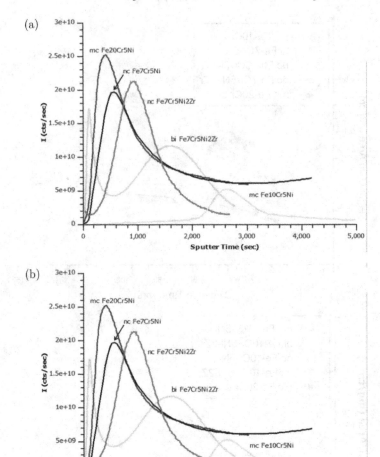

Fig. 6. SIMS depth profiles showing enrichment of Cr in the oxide layer of Fe–Cr–Ni–Zr alloys having nc and bimodal grain size distributions, compared with mc high-Cr Fe20Cr5Ni and mc Fe10Cr5Ni alloys (a) 7Cr and (b) 10Cr alloys.[97]

oxidation resistance). A plausible mechanism for such a protective layer formation based on enhanced migration of Cr in finer grains and lateral growth of chromia into a full-fledged continuous film has been proposed.[97,98]

Dispersion of coarse grain in a nc matrix is an efficient way for improving the ductility of nc alloys without losing the benefits of greater solute migration and high oxidation resistance. Although a few studies have been conducted on electrochemical corrosion behavior of bimodal alloys,[99,100] the results shown in Figs. 5 and 6 were the first on high temperature oxidation resistance of such alloys. Bimodal alloys present an inexpensive alternative to the currently used high temperature alloys.

6. Summary

As a result of their extremely fine grain size, nc metals and alloys possess remarkably different properties. However, because such materials have poor thermal stability and are highly susceptible to grain growth, their large scale processing is a challenge. But recent research has helped to develop a sound understanding of the grain coarsening process in nc materials and ways to address the problem of grain growth.

Alloys containing elements that can form protective oxide scale (namely, Cr, Si, and Al) have shown superior resistance to degradation due to gaseous corrosion at elevated temperatures. Such alloys are attractive in both bulk and coatings forms for various industrial applications including automotive, aerospace, power generation, structural etc.

Recent studies have established the nc structure to possess a remarkably superior resistance to oxidation of an Fe–10Cr alloy (than its mc counterpart) at moderately high temperatures (300–400°C). This behavior has been convincingly attributed to the much greater diffusivity of chromium in the nc alloy, and as a result, the ability of the nc alloy to develop a protective layer of chromium oxide (in spite of only 10 wt.% Cr in the alloy).

Lower ductility and formability of nc alloys are the biggest impediment in their commercial-scale applications. However, use of nc alloys as coatings or as bimodal alloys is a plausible option for exploiting the incredible properties inherent to nc materials.

References

1. R.W. Siegel, Nanostructured materials — mind over matter, *Nanostruct. Mater.* **4**(1) (1994), pp. 121–138.
2. C. Suryanarayana and F. Froes, The structure and mechanical properties of metallic nanocrystals, *Metall. Mater. Trans. A* **23**(4) (1992), pp. 1071–1081.
3. H.E. Schaefer *et al.*, Structure and properties of nanometer-sized solids, in *Energy Pulse and Particle Beam Modification of Materials*, K. Henning (Ed.) (Akademie-Verlag, Berlin, 1988).
4. C.C. Koch *et al.*, Breakthroughs in optimization of mechanical properties of nanostructured metals and alloys, *Adv. Eng. Mater.* **7**(9) (2005), pp. 787–794.
5. R.K. Singh Raman, R.K. Gupta and C.C. Koch, Resistance of nanocrystalline *vis-à-vis* microcrystalline Fe–Cr alloys to environmental degradation and challenges to their synthesis, *Phil. Mag. A* **90**(23) (2010), pp. 3233–3260.
6. R. Rofagha *et al.*, The effects of grain size and phosphorus on the corrosion of nanocrystalline Ni-P alloys, *Nanostruct. Mater.* **2**(1) (1993), pp. 1–10.
7. R. Kirchheim *et al.*, Free energy of active atoms in grain boundaries of nanocrystalline copper, nickel and palladium, *Nanostruct. Mater.* **1**(2) (1992), pp. 167–172.
8. K.E. Knipling, M. Daniil and M.A. Willard, Fe-based nanocrystalline soft magnetic alloys for high-temperature applications, *Appl. Phys. Lett.* **95**(22) (2009).
9. M.A. Willard *et al.*, Soft magnetic nanocrystalline alloys for high temperature applications, 2002, DTIC Document.
10. D. Sebayang *et al.*, *Key Eng. Mater.* **474–476** (2011), pp. 2134–2139.
11. Z. Huang *et al.*, *J. Elec. Soc.* **156**(3) (2009), pp. C95–C102.
12. T. Moskalewicz, B. Dubiel and B. Wendler, AlCuFe(Cr) and AlCoFeCr coatings for improvement of elevated temperature oxidation resistance of a near-α titanium alloy, *Mater. Charact.* **83** (2013), pp. 161–169.
13. B. Wendler, L. Kaczmarek and L. Klimek, *Oxidation Resistance of nanocrystalline microalloyed TiAl.*, *13th International scientific conference on achievements in mechanical and materials engineering*, Poland, 2005.
14. T. Chookajorn, H.A. Murdoch and C.A. Schuh, Design of stable nanocrystalline alloys, *Science* **337**(6097) (2012), pp. 951–954.
15. H. Gleiter, Nanocrystalline materials, *Prog. Mater. Sci.* **33**(4) (1989), pp. 223–315.
16. R.W. Siegel, Nanostructured materials — mind over matter, *Nanostruct. Mater.* **3**(1–6) (1993), pp. 1–18.

17. R.W. Siegel, *Cluster-assembled nanophase materials, Annu. Rev. Mater. Sci.* **21**(1) (1991), pp. 559–578.
18. R.W. Siegel, *Synthesis and properties of nanophase materials, Mater. Sci. Eng. A* **168**(2) (1993), pp. 189–197.
19. R.W. Siegel, Nanophase Materials Assembled from Atomic Clusters, *MRS Bulletin* **15**(10) (1990).
20. G.J. Thomas, R.W. Siegel, and J.A. Eastman, Grain boundaries in nanophase palladium: High resolution electron microscopy and image simulation, *Scripta. Metall. Mater.* **24**(1) (1990), pp. 201–206.
21. X. Zhu *et al.*, X-ray diffraction studies of the structure of nanometer-sized crystalline materials, *Phys. Rev. B* **35**(17) (1987), p. 9085.
22. T. Haubold *et al.*, Exafs studies of nanocrystalline materials exhibiting a new solid state structure with randomly arranged atoms, *Phys. Lett. A* **135**(8–9) (1989), pp. 461–466.
23. T. Mutschele and R. Kirchheim, Hydrogen as a probe for the average thickness of a grain boundary, *Scr. Metall.* **21**(8) (1987), pp. 1101–1104.
24. J. Horvath, R. Birringer and H. Gleiter, Diffusion in nanocrystalline material, *Solid State Commun.* **62**(5) 1987, pp. 319–322.
25. G. Wallner *et al.*, *Mater. Res. Symp. Proc.* **132** (1989), p. 149.
26. H. Gleiter, Nanostructured materials: Basic concepts and microstructure, *Acta Mater.* **48**(1) (2000), pp. 1–29.
27. C. Suryanarayana, Nanocrystalline materials, *Int. Mater. Rev.* **40** (1995), pp. 41–64.
28. R. Birringer, Nanocrystalline materials, *Mater. Sci. Eng. A* **117** (1989), pp. 33–43.
29. K. Lu, Nanocrystalline metals crystallized from amorphous solids: Nanocrystallization, structure, and properties, *Mater. Sci. Eng. R* **16** (1996), p. 61.
30. T.R. Malow and C.C. Koch, Mechanical properties in tension of mechanically attrited nanocrystalline iron by the use of the miniaturized disk bend test, *Acta Mater.* **46**(18) (1998), pp. 6459–6473.
31. J. Joardar *et al.*, Stability of nanocrystalline disordered NiAl synthesized by mechanical alloying, *Philos. Mag. A* **82** (2002), pp. 469–475.
32. H. Natter *et al.*, *In Situ* X-ray crystallite growth study on nanocrystalline Fe, *Mater. Sci. Forum* **343–346** (2000).
33. H. Natter *et al.*, Grain-growth kinetics of nanocrystalline iron studied *in situ* by synchrotron real-time x-ray diffraction, *J. Phys. Chem. B* **104**(11) (2000), pp. 2467–2476.
34. R. Perez *et al.*, Grain growth of nanocrystalline cryomilled Fe–Al powders, *Metall. Mater. Trans. A* **29**(10) (1998), pp. 2469–2475.

35. R.J. Perez, H.G. Jiang and E.J. Lavernia, Grain size stability of nanocrystalline cryomilled Fe-3wt%Al alloy, *Nanostruct. Mater.* **9**(1–8) (1997), pp. 71–74.

36. E. Bonetti *et al.*, Thermal evolution of ball milled nanocrystalline iron, *Nanostruct. Mater.* **12**(5–8) (1999), pp. 685–688.

37. C.H. Moelle and H.J. Fecht, Thermal stability of nanocrystalline iron prepared by mechanical attrition, *Nanostruct. Mater.* **6**(1–4) (1995), pp. 421–424.

38. A. Michels *et al.*, Modelling the influence of grain-size-dependent solute drag on the kinetics of grain growth in nanocrystalline materials, *Acta Mater. A* **47**(7) 1999, pp. 2143–2152.

39. P.C. Millett, R.P. Selvam and A. Saxena, Molecular dynamics simulations of grain size stabilization in nanocrystalline materials by addition of dopants, *Acta Mater.* **54** (2006), pp. 297–303.

40. J. Weissmuller, Alloy effects in nanostructures, *Nanostruct. Mater.* **3** (1993), pp. 261–272.

41. P. Wynblatt and R.C. Ku, Surface energy and solute strain energy effects in surface segregation, *Surf. Sci.* **65** (1977), pp. 511–531.

42. H. Tanimoto *et al.*, Self-diffusion in high-density nanocrystalline Fe, *Nanostruct. Mater.* **12**(5–8) (1999), pp. 681–684.

43. A. Hernando *et al.*, *Electronic Transport in Nanocrystalline Iron: A Low T Magnetoresistance Effect* (Elsevier, California, 2003).

44. G.E. Fougere, J.R. Weertman and R.W. Siegel, Processing and mechanical behavior of nanocrystalline Fe, *Nanostruct. Mater.* **5**(2) (1995), pp. 127–134.

45. D. Segers *et al.*, Positron annihilation study of nanocrystalline iron, *Nanostruct. Mater.* **12**(5–8) (1999), pp. 1059–1062.

46. A.A. Karimpoor *et al.*, High strength nanocrystalline cobalt with high tensile ductility, *Scripta. Mater.* **49**(7) (2003), pp. 651–656.

47. C. Wang *et al.*, Review on modified and novel techniques of severe plastic deformation. Science, *China Technological Sciences*, **55**(9) (2012), pp. 2377–2390.

48. R.Z. Valiev, Nanostructuring of metals by severe plastic deformation for advanced properties, *Nat. Mater.* **3** (2004), pp. 511–516.

49. R.Z. Valiev, R.K. Islamgaliev and I.V. Alexandrov, Bulk nanostructured materials from severe plastic deformation, *Prog. Mater. Sci.* **45** (2000), p. 103.

50. G.D. Hibbard, K.T. Aust and U. Erb, Thermal stability of electrodeposited nanocrystalline Ni-Co alloys, *Mater. Sci. Eng. A* **433**(1–2) (2006), pp. 195–202.

51. J.H. Seo *et al.*, Textures and Grain Growth in Nanocrystalline Fe-Ni Alloys, *Mater. Sci. Forum* (2005), pp. 475–479.

52. J.R. Groza, *Nanocrystalline Powder Consolidation Methods,* in *Nanostructured Materials: Processing, Properties, and Applications,* C.C. Koch, (Ed.) (William Andrew Pub, Norwich, NY, 2007).

53. K.M. Youssef *et al.*, Ultratough nanocrystalline copper with a narrow grain size distribution. *Appl. Phys. Lett.* **85**(6) (2004), pp. 929–931.

54. O. Elkedim, H.S. Cao and D. Guay, Preparation and corrosion behavior of nanocrystalline iron gradient materials produced by powder processing, *J. Mater. Process. Tech.* **121**(2–3) (2002), pp. 383–389.

55. S.K. Vajpai, B.V. Mahesh, and R.K. Dube, Studies on the bulk nanocrystalline Ni–Fe–Co alloy prepared by mechanical alloying–sintering–hot rolling route, *J. Alloys Compd.* **476**(1–2) (2009), pp. 311–317.

56. S.K. Vajpai, R.K. Dube and A. Tewari, Studies on the mechanical alloying of Ni–Fe–Co powders and its explosive compaction, *Metall. Mater. Trans. A* **39**(11) (2008), pp. 2725–2735.

57. A.P. Zhilyaev *et al.*, Microhardness and microstructural evolution in pure nickel during high-pressure torsion, *Scripta. Mater.* **44** (2001), pp. 2753–2758.

58. A.P. Zhilyaev *et al.*, Experimental parameters influencing grain refinement and microstructural evolution during high-pressure torsion, *Acta Mater.* **51** (2003), pp. 753–765.

59. Y. Iwahashi *et al.*, The process of grain refinement in equal-channel angular pressing, *Acta Mater.* **46** (1998), pp. 3317–3331.

60. Y. Iwahashi *et al.*, Principle of equal-channel angular pressing for the processing of ultra-fine grained materials, *Scripta. Mater.* **35** (1996), pp. 143–146.

61. L.J. Kecskes *et al.*, ECAE processing of pure and mg alloy powders: Effect of confinement, route, and temperature, in *Ultra fine Grained Materials Sixth International Symposium,* TMS Annual Meeting and Exhibition, 2010.

62. M. Jain and T. Christman, Synthesis, processing, and deformation of bulk nanophase Fe–28Al–2Cr intermetallic, *Acta Metall. Mater.* **42**(6) (1994), pp. 1901–1911.

63. L. Fu *et al.*, Nanostructured hypoeutectic FeB alloy prepared by a self propagating high temperature synthesis combining a rapid cooling technique, *Nanoscale Res. Lett.* **4** (2008), pp. 11–16.

64. G.-G. Lee and S.-D. Kim, Arc plasma synthesis of nanostructured Fe powder, *Mater. Res. Symp. Proc.* **740** (2003), pp. 13–22.

65. T. Saito, T. Takeuchi and H. Kageyama, *J. Mater. Res.* **19**(9) (2004), pp. 2730.

66. P. Kofstad, *High Temperature Corrosion.* Vol. 6 (Elsevier Applied Science and Publishers Ltd, New York, 1988).

67. H.Y. Tong, F.G. Shi, and E.J. Lavernia, Enhanced oxidation resistance of nanocrystalline FeBSi materials, *Scripta. Metall. Mater.* **32**(4) (1995), pp. 511–516.

68. O. El Kedim *et al.*, Electrochemical behavior of nanocrystalline iron aluminide obtained by mechanically activated field activated pressure assisted synthesis, *Mater. Sci. Eng. A* **369** (1–2) (2004), pp. 49–55.

69. Z.B. Wang *et al.*, Diffusion of chromium in nanocrystalline iron produced by means of surface mechanical attrition treatment, *Acta Mater.* **51**(14) (2003), pp. 4319–4329.

70. R.K. Singh Raman and R.K. Gupta, Oxidation resistance of nanocrystalline vis-à-vis microcrystalline Fe–Cr alloys, *Corros. Sci.* **51**(2) (2009), pp. 316–321.

71. Y. Shida *et al.*, JIMS-3, *High Temp. Corros.*, *T. Jpn. I. Met.* 1983.

72. R.K. Singh Raman, Influence of microstructural variations in the weldment on the high-temperature corrosion of 2.25Cr-1Mo steel, *Metall. Mater. Trans. A* **26**(7) (1995), pp. 1847–1858.

73. R.K. Singh Raman, Role of microstructural degradation in the heat-affected zone of 2.25Cr–1Mo steel weldments on subscale features during steam oxidation and their role in weld failures, *Metall. Mater. Trans. A* **28** (1998), p. 577.

74. R.K. Singh Raman, J.B. Gnanamoorthy and S.K. Roy, Oxidation behavior of 2.25Cr–1Mo steel with prior tempering at different temperatures, *T. Indian I. Metals* **46** (1992), p. 391

75. R.K. Singh Raman, A.S. Khanna, and J.B. Gnanamoorthy, *J. Mater. Sci. Lett.* **9** (1990), p. 353.

76. R. Gupta, R.K. Singh Raman and C.C. Koch, Grain growth behaviour and consolidation of ball-milled nanocrystalline Fe-10Cr alloy, *Mater. Sci. Eng. A*, **494**(1–2) (2008), pp. 253–256.

77. C. Koch, Ductility in nanostructured and ultra fine-grained materials: Recent evidence for optimism, *J. Metastable and Nanocrys. Mater.* **18** (2003), p. 9.

78. C. Koch *et al.*, Ductility of nanostructured materials, *MRS Bulletin* **24** (1999), p. 54.

79. H.V. Swygenhoven and A. Caro, Molecular dynamics computer simulation of nanophase Ni: structure and mechanical properties, *Nanostruct. Mater.* **9** (1997), p. 669.

80. J.E. Carlsley *et al.*, Mechanical behavior of a bulk nanostructured iron alloy, **29** (1998), pp. 2261–71.

81. R.S. Iyer *et al.*, Plastic deformation of nanocrystalline Cu and Cu-0.2 wt% B, *Mater. Sci. Eng. A*, **264** (1999), p. 210.

82. S.C. Tjong and H. Chen, Nanocrystalline materials and coatings, *Mat. Sci. Eng. R* **45** (2004) p. 1–88.

83. F. Wang and S. Geng, *Surf. Eng.* **19**(1) (2003), p. 32.

84. H. Lou *et al.*, Oxidation behavior of sputtered microcrystalline coating of superalloy K17F at high temperature, *Mater. Sci. Eng. A* **207** (1996), p. 121.

85. H. Lou *et al.*, High-temperature oxidation resistance of sputtered micro-grain superalloy K38G, *Oxid. Met.* **38** (1992), p. 299.

86. H. Lou *et al.*, Oxide formation of K38G superalloy and its sputtered micrograined coating, *Surf. Coat. Tech.* **63** (1994), p. 105.

87. J. Zhang and H. Lou, *Corros. Sci. Prot. Tech.* **10** (1998), p. 82.

88. J.X. Zhang and H. Lou, Oxidation behaviour of the sputtered microcrystalline coating of LDZ125 superalloy at high temperatures, *Acta Metall. Sin.* **34** (1998), p. 627.

89. F. Wang, *High Tempereature Nanocrystalline Coatings* (The Chinese Academy of Sciences, 1992).

90. Z. Liu *et al.*, Oxidation behaviour of sputter-deposited Ni–Cr–Al micro-crystalline coatings, *Acta Mater.* **46** (1998), p. 1691.

91. Z. Liu *et al.*, The effect of coating grain size on the selective oxidation behaviour of Ni–Cr–Al alloy, *Scripta. Mater.* **37** (1997), p. 1551.

92. Z. Liu *et al.*, *Scripta. Mater.* **37** (2002), p. 1151.

93. F.H. Stott, G.C. Wood and J. Stringer, The influence of alloying elements on the development and maintenance of protective scales, *Oxid. Met.* **44** (1995), p. 113.

94. G.F. Chen and H. Lou, Effect of nanocrystallization on the oxidation behavior of a Ni–8Cr–3.5Al alloy, *Oxid. Met.* **54**(1/2) (2000), p. 155.

95. Y.M. Wang and E. Ma, Three strategies to achieve uniform tensile deformation in a nanostructured metal, *Acta Mater.* **52**(6) (2004), pp. 1699–1709.

96. B.V. Mahesh, R.K. Singh Raman and C.C. Koch, Bimodal grain size distribution: An effective approach for improving the mechanical and corrosion properties of Fe–Cr–Ni alloys, *J. Mater. Sci.* **47**(22) (2012), pp. 7735–7743.

97. B.V. Mahesh *et al.*, Fe–Cr–Ni–Zr alloys with bi-modal grain size distribution: Synthesis, Mechanical properties and oxidation resistance, *Mater. Sci. Eng. A* **574** (2013), pp. 235–242.

98. B.V. Mahesh, Nanocrystalline and bimodal fe-cr alloys: synthesis, mechanical peoperties and oxidation resistance, Department of Mechanical Engineering, Monash University, Melbourne, 2013

99. S. Gollapudi, Grain size distribution effects on the corrosion behaviour of materials, *Corrosion Sci.* **62**(0) (2012), pp. 90–94.

100. C.D. Gu *et al.*, Electrodeposition, structural, and corrosion properties of Cu films from a stable deep eutectics system with additive of ethylene diamine, *Surf. Coat. Technol.* **209** (2012), pp. 117–123.

Chapter 10

Reactive Element Additions in High Temperature Alloys and Coatings

D. Naumenko and W.J. Quadakkers

Institute for Energy and Climate Research (IEK-2)
Forschungszentrum Jülich GmbH 52425, Jülich, Germany

1. Introduction

First mentioned in the patent of Pfeil of 1937,[1] minor (typically 0.01–0.5 wt.%) additions of reactive elements (REs) from Groups IIIA to IVA, e.g., Y, La, Zr, etc., are common constituents of the state-of-the-art high temperature metallic alloys and coatings. The RE additions, either in metallic form or as oxide dispersions of respective metals, significantly improve the adherence of alumina and chromia based surface scales. Moreover, REs reduce the growth rate of both oxide scale types, the extent of reduction being much more pronounced in chromia than in alumina scales. Better adherence and reduced growth kinetics of oxide scales associated with RE-additions lead to considerable extension of lifetime of alloy components and coatings limited by consumption of Al and/or Cr reservoir as a result of scale formation and rehealing in the case of spallation.[2] The oxidation resistance increase by RE additions justifies higher material costs related to technological challenges of adding precisely controlled, small amounts of highly reactive metals in large alloy/coating production facilities.

In the following, the mechanisms of the RE-effect will be briefly discussed, emphasizing the progress made in the last 30 years mainly related to advancement of analytical techniques. Furthermore, the practically relevant aspects of RE-additions in alloy/coating systems will be elucidated mainly using alumina forming materials as examples.

2. The Mechanisms of the Reactive Element Effect

2.1 *General Remarks*

Numerous papers exist in which the mechanisms which are responsible for the improved oxidation properties due to RE additions are discussed (see e.g., review papers in references).[3-5] Many mechanisms originally proposed to explain the reactive element effect appeared to be not valid or play only a minor role. For instance, formation of RE-containing oxide pegs[6] was claimed to be responsible for the improvement of oxide adhesion, by "anchoring" the scale to the substrate. However, in later studies many alloy systems such as ODS FeCrAl-alloys were found to form adherent scales of uniform thickness, so the pegging is, at least in these alloys, likely to be of secondary importance. Alternatively, provision of vacancy sinks by the RE-oxide particles in the FeCrAl ODS alloys was suggested to effectively prevent their condensation into voids at the scale/metal interface.[3] Considering later oxygen tracer experiments revealing that the alumina scales on RE-containing alloys grow by inward oxygen diffusion (see Section 2.2 and references),[7,8] this mechanism appears to be hardly plausible.

    The majority of studies seem to confirm that two dominating RE-mechanisms are change of the oxide scale transport properties due to "decoration" of scale grain boundaries by REs[9] and gettering of alloy impurities such as S.[10-12]

2.2 *Effect of REs on Alumina Scale Microstructure and Transport Processes*

It is generally accepted that minor additions of the REs change the transport phenomena on the grain boundaries of the oxide scales. It is doubtful whether the RE-additions can substantially influence the diffusion through the bulk alpha-Al_2O_3, e.g., by acting as donors and/or acceptors, thereby changing the defect concentrations in the oxide. Recent studies of tracer distribution and electrical conductivity of RE-doped alumina single-crystals have shown only minor, if any, effect on lattice diffusion.[13,14] Moreover, the diffusion coefficients of aluminum and oxygen determined experimentally in bulk alumina in the temperature range of interest, i.e., 900–1300°C are orders of magnitude lower than those at the grain boundaries of polycrystalline alumina.[15,16]

The evidence of RE-effects on the diffusion processes at the alumina grain boundaries have been delivered by two-stage oxidation experiments using oxygen isotopes.[7,8,17,18] For this purpose, a material is e.g., exposed first in an ^{18}O enriched atmosphere and then the gas supply is switched at oxidation temperature to an atmosphere with normal ^{16}O. The isotope distribution in the formed oxide scale is subsequently analyzed using secondary ion mass spectrometry (SIMS) or sputtered neutrals mass spectrometry (SNMS).[7,8] Typical isotope distributions in the oxide scales formed on alumina forming alloys with and without RE-additions are shown in Fig. 1. Qualitatively similar RE-effects on the isotope distribution were obtained with chromia forming materials.[8] Analysis of Fig. 1 reveals that the alumina scales on FeCrAl alloys without RE-additions grow by both inward oxygen and outward aluminum grain boundary transport (enrichment by second isotope at both scale interfaces). In contrast, the scales on the RE-containing alloys formed almost exclusively by inward oxygen grain boundary diffusion (major peak only at the scale/metal interface). The RE-alloying additions apparently suppress the outward transport of aluminum cations.[8]

Extensive secondary and transmission electron microscopy (TEM) SEM studies revealed the oxide scales on RE-free alloys to exhibit an equiaxed grain structure, whereas those on RE-containing alloys feature columnar grains with the lateral grain size increasing into growth direction (toward the scale/metal interface; Fig. 2). Using high resolution TEM/EDX analyses, segregation of the REs and in some cases formation of Y-containing compounds at the

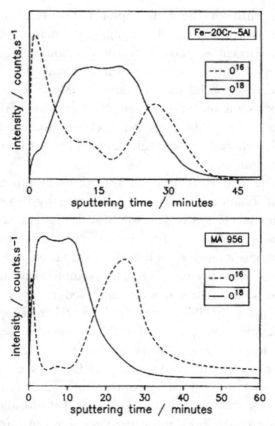

Fig. 1. Oxygen isotope distribution measured by SIMS in oxide scales formed on Fe-20Cr-5Al alloy (a) and Fe-20Cr-4.5Al-0.5Y$_2$O$_3$-0.4Ti (ODS-Alloy MA956) after 3 h in ^{18}O-enriched air and subsequently 9 h in non-enriched air at 1000°C.[8]

scale grain boundaries and at the scale/metal interface was verified (Fig. 3).[19-21,23]

Apart from decorating the scale grain boundaries, the reactive element yttrium has been found to diffuse through the alumina scale to the scale/gas interface and to form there precipitates of mixed oxides such as Y$_3$Al$_5$O$_{12}$.[9] These precipitates were found to grow with time, indicating a continuous diffusion process of yttrium through the scale.[24] Pint[9] related the RE-effects to experimental scanning TEM (STEM) data of RE-ions segregation to the scale boundaries and to the scale/metal interface. The proposed "dynamic segregation

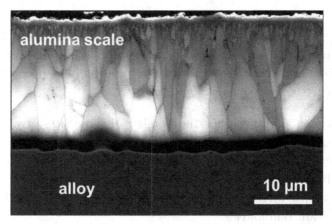

Fig. 2. SEM In-Lens image of alumina scale formed on Fe-20Cr-5Al-0.05Y model wrought alloy after 2000 h discontinuous oxidation at 1200°C in laboratory air.[22]

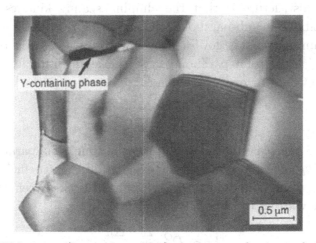

Fig. 3. TEM picture (transverse section) of columnar alumina scale formed on FeCrAl ODS alloy MA 956 after 1000 h oxidation at 1200°C.[23]

theory" claims that the RE-ions first segregate to the scale/metal interface and then driven by the oxygen chemical potential gradient, diffuse outwards through the scale. During this outward diffusion, the RE-ions block outward transport of Al-cations at the scale grain boundaries.

The driving force for the segregation of REs to the alumina grain boundaries and the oxide/metal interface has been claimed to be the misfit energy of the large RE-ions in the bulk alumina.[9] Although this is a plausible assumption, the exact way, in which the REs block the cation transport on the alumina grain boundaries is not completely understood and several alternative mechanisms have been proposed in literature:[7,19,25–27]

(1) REs may simply reduce the number of ionic sites, available for diffusion of Al cations;
(2) They may change the boundary atomic structure to suppress Al-diffusion;
(3) They may increase bonding strength between Al and O ions at the grain boundaries.[28]

One of the consequences of the reactive element effect on the scale microstructure is that the alumina scaling kinetics changes from parabolic to sub-parabolic.[27] Considering that the growth rate of the alumina scale is governed by inward oxygen grain boundary diffusion,[22] the usual parabolic scale thickening rate for constant grain size would be expected, namely:

$$\frac{dx}{dt} = \frac{k_P}{x}, \tag{1}$$

where x is the scale thickness and t is the time. The parabolic rate constant k_P is proportional to the oxygen diffusion coefficient along the scale grain boundaries D_{GB}^0 and the ratio between the grain boundary width δ to the grain size r:[22]

$$k_P \sim D_{GB}^O \frac{2\delta}{r} \cdot \frac{\Delta\mu}{RT}, \tag{2}$$

where $\Delta\mu$ is the oxygen potential gradient.

The experimentally observed microstructure of the alumina scales grown on RE-doped alloys features columnar grains whereby the grain width r can be described by:

$$r(x) = r_0 + a \cdot x, \tag{3}$$

where r_0 is the oxide grain width at the scale/gas interface and a is a proportionality factor.[22] Assuming that the oxygen diffusion through the scale is limited by the areas with the largest oxide grain size, i.e., at the scale/metal interface, Equations (1)–(3) can be combined to deliver after integration an expression to describe the experimentally observed sub-parabolic oxidation kinetics:

$$\frac{ax^3}{3} + \frac{r_0 x^2}{2} = 2D_{GB}^O \delta_{GB} \frac{\Delta\mu}{RT} t. \tag{4}$$

The implication of Equation (4) is that in the initial stages of oxidation, i.e., at small values of x a nearly parabolic oxidation rate is observed. Upon scale thickening, the first term on the left hand side of Equation (4) starts to dominate and the oxidation rate becomes essentially cubic. Recently, it was proposed that Equation (4) has to be slightly modified if a different oxygen activity gradient within the scale prevails, i.e., if the oxygen diffusion in the outer part of the scale has to be accounted for.[29]

2.3 *Effect of REs on Oxide Scale Adherence*

In the context of the RE-effect on oxide scale adherence, it is necessary to consider the deleterious effect of sulfur, which has been known for around three decades.[30] The presence of sulfur at the scale/metal interface of undoped FeCrAl alloys was experimentally verified by AES studies after different oxidation times[31] and associated with interfacial voids.[11] However, it is still a point of debate whether the sulfur segregates to the intrinsically strong oxide/metal interface[11] or whether it enhances the growth of existing voids at this interface.[32] Sulfur containing, reactive element free alloys are prone to early and severe scale spallation. Hydrogen annealing of alumina formers reduces the bulk S-content resulting in substantial improvement of the alumina scale adherence.[31] The alloy manufacturers are well aware of the negative role of sulfur and its content in commercial high temperature materials such as FeCrAl is therefore commonly kept below 20 wt. ppm.

It has been suggested[30] that the REs form stable sulfides in the alloy matrix, thus preventing sulfur from segregating to the oxide/metal interface. The sulfide formation in the bulk alloy (NiAl) was verified by high resolution TEM analysis.[33]

Using theoretical considerations[34] it was proposed that the segregation of RE-atoms or ions to the alumina/metal interface can enhance the scale adhesion and/or prevent sulfur segregation. The latter argument is indirectly supported by the observation that RE-alloying provides better scale adherence compared to sulfur removal by hydrogen annealing.[31]

Finally, the change in the scale growth mechanism on RE-doped alloys described in Section 2.1 can by itself explain the improvement of oxide scale adherence as suggested in.[27] The cation flux in RE-free alloys should result in an opposite flux of vacancies, which might condense at the scale/metal interface promoting delaminations and voids thereby decreasing the oxide scale adherence. In RE-doped scales, the inward oxygen ion flux should effectively remove vacancies from the scale/metal interface. Moreover, the scale adherence of RE-doped alloys (especially chromia formers) may benefit from a decrease in the oxide thickening rate and especially from reduction of the compressive oxide growth stress.[27] The latter was commonly observed to result in convoluted scale morphologies in the case of undoped alloys.[4]

3. Selected Aspects of REs Use in Practical Applications

3.1 *General Remarks*

The property improvement which can be obtained by addition of REs, such as Y, La, and Ce to chromia and alumina forming alloys depends on the type of RE used,[35-37] its amount, distribution and the method by which it is introduced into the alloy (e.g., as metallic addition or in form of an oxide dispersion) or to the alloy surface (e.g., by implantation, dipping, or sol–gel techniques). No general simple rule exists from which the optimum type and amount of a reactive element can be derived, i.e., the addition(s) to be used depend(s) on the actual alloy system and the way in which the REs interact with major and minor alloying additions and impurities. In the following, a number of examples of the practically relevant aspects of RE-additions will be briefly discussed.

For detailed descriptions, the reader is referred to the respective publications.

3.2 Chromia Forming Oxide Dispersion Strengthened (ODS) and Wrought Alloys

In ODS alloys, the size and distribution of the RE-dispersions appeared to have a substantial effect on the oxidation rate. Cr5Fe-based alloys manufactured by mechanical alloying (MA; dispersion size of 10 nm, inter-particle distance of 100 nm) showed a lower oxidation rate than those produced by elemental mixing (MIX; dispersion size of 1 μm, inter-particle distance of 5 μm), as shown in Fig. 4.[38,39] Interestingly, improvement in oxide scale adherence by the yttria dispersion compared to the RE-free alloy occurs in both alloys, i.e., irrespective of the exact dispersion size and distribution. For obtaining a decrease in oxide growth rate the dispersion distribution appeared to be of more importance than the type (Y_2O_3, La_2O_3, CeO_2; Fig. 5) or exact amount (0.2–1 wt.%) of the actually prevailing reactive element dispersion.

Figure 6 shows gravimetric data of Fe-24Cr alloys with addition of different REs during oxidation at 800°C. All RE-additions lead

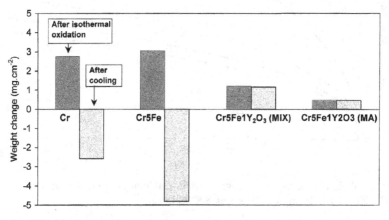

Fig. 4. Weight change data of chromium base alloy Cr5Fe without and with 1 wt.-% yttria (MA and MIX) after 100 h isothermal oxidation at 1000°C in Ar/O_2 compared with weight losses during subsequent cooling to room temperature.[39]

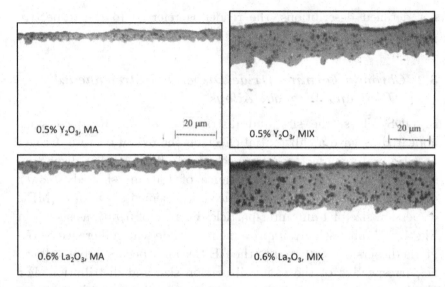

Fig. 5. Surface scales on various Cr5Fe-based ODS-alloys after 1000 h oxidation at 950°C in $Ar/H_2/H_2O$.[39]

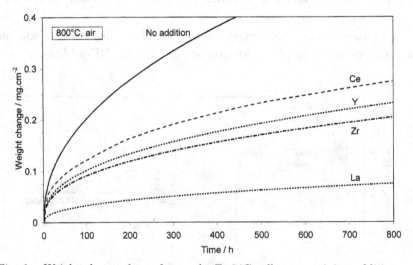

Fig. 6. Weight change data of wrought Fe-24Cr alloys containing additions of Y, Ce, Zr, or La during oxidation at 800°C in air.[39]

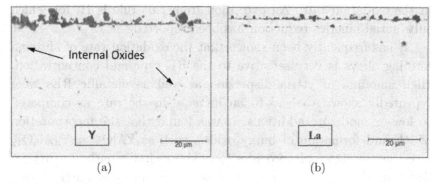

Fig. 7. Metallographic cross sections of selected alloys in Fig. 6 showing scale formation and different morphologies of internal oxidation after 1000 h exposure at 800°C.[39]

to a decrease in oxide growth rate, but apparently the La addition is in this respect more effective compared to Y, Ce, or Zr. The metallographic cross-sections in Fig. 7 provide an explanation for this finding. Y, Zr, and Ce possess hardly any solubility in the Fe-matrix and therefore form Fe-rich intermetallic compounds (e.g., $Fe_{17}Ce_2$, $Fe_{17}Y_2$). These intermetallics tend to precipitate at the alloy grain boundaries and, upon high temperature exposure, they are transformed into RE-rich internal oxides beneath the surface scale. This results in a non-even distribution of RE, thus hampering rapid, homogeneous incorporation in the chromia scale. In the case of La, this effect does not occur because La does not form intermetallic compounds with Fe and, at 800°C, it can be dissolved in the α-Fe matrix up to a concentration of around 0.5%. The La will therefore, after being transformed into internal oxides, be much more evenly distributed in the alloy, and become more easily incorporated into the chromia scale than Y, Ce, or Zr.

3.3 *Reactive Element Reservoir in FeCrAl-Alloys*

Contrary to chromia forming alloys, incorporation of REs into alumina scales in most cases does not result in a drastic decrease in the oxidation rate. An exception from this rule is Hf for which substantial kinetics reduction has been reported.[40]

It has frequently been shown that the oxidation rate of alumina forming alloys is very sensitive to the RE type and concentration. High amounts of yttria dispersion as well as metallic REs were repeatedly shown to lead to an increase in the rate as compared to low or moderate additions. It was found that the incorporation of RE and formation of bulky oxides, such as $YAlO_3$ or $Y_3Al_5O_{12}$ into the alumina scale provides local short circuit paths for oxygen diffusion, thereby increasing the scale thickening rate. Furthermore, extensive internal oxidation of the REs is observed when they are added in metallic form. The faster oxidation rate associated with high RE-contents is commonly termed "overdoping".[41] Therefore, many research efforts have been spent in optimizing the RE-content with respect to reduction of the oxidation rate for various alloy types.[42]

In practice, the RE-content optimisation appeared to be difficult due to the fact that the RE-effect on the scale growth rate and adherence is a matter of RE-reservoir.[43,44] In addition to the RE-alloy content, the RE-reservoir appeared to be determined by the component surface to volume ratio (wall thickness) as well as the contents of alloy impurities, which are introduced during alloy manufacturing and which may tie-up the REs within the bulk alloy thereby preventing their incorporation into the oxide scale. The concept of RE-reservoir was originally introduced in a study of a high-purity Zr-containing FeCrAlY alloy.[43] It was found (Fig. 8) that the oxidation rate of a 0.3 mm thick specimen was significantly lower than that of a 1.0 mm thick sample of the same material.

Microstructural studies of the FeCrAlYZr alloy revealed that the enhanced oxidation rate on the thicker sample could be correlated with its higher Zr-reservoir. The latter caused more extensive incorporation of zirconia particles into the alumina scale thus providing short circuit paths for oxygen diffusion. After Zr had been depleted from the alloy, the alumina scale growth rate decreased (Fig. 9) and the microstructure of the newly formed oxide scale became similar

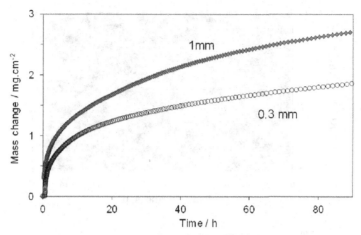

Fig. 8. Thermogravimetrical data of wrought Fe-20Cr-5Al-0.05Y-0.03Zr (wt.%) model alloy of two different specimen thicknesses during isothermal oxidation at 1300°C in Ar-20%O$_2$.[43]

to that of the Zr-free material, i.e., forming large, columnar alumina grains near the scale/alloy interface.

3.4 *Interaction of REs with C and N Impurities*

The importance of RE interaction with alloy impurities of C and N has been illustrated by the effect of minor additions of IVa-group elements (Ti, Zr, and Hf) on the scale spallation resistance of FeCrAlY-type alloy. In a IVa-elements free but Y-containing model alloy, the presence of Cr-carbide at the scale/metal interface led to considerable scale spallation after long time exposure.[45] Although not as severe as in the case of RE-free alloys the latter effect appeared to substantially shorten the oxidation limited lifetime caused by Al-consumption as a result of scale re-healing after repeated spalling. In contrast, additions of IVa elements in the amount of 0.03–0.05 wt.% effectively prevented the interfacial Cr-carbide precipitation by forming thermodynamically more stable carbides. As a result, the lifetimes of the alloys containing additions of IVa elements were substantially extended compared to the one without such addition (Fig. 10).

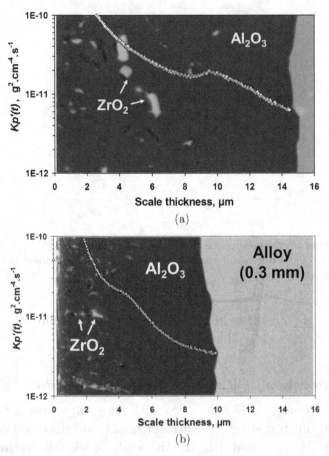

Fig. 9. SEM cross-section of model Fe-20Cr-5Al-0.05Y-0.03Zr alloy after 96 h isothermal oxidation at 1300°C in Ar-20%O_2 superimposed with the plot of "apparent" instantaneous parabolic rate constant $K'_P(t)$ derived from the mass change data in Fig. 8 for specimen with 1mm thickness (a) and 0.3 mm thickness (b).[40,43]

The exact effect on the alumina scale microstructure, growth rate and adherence strongly depends on the form and distribution in which IVa elements are present in the alloy. In alumina forming alloys with high C and/or N contents, the coarse precipitates of Ti, Zr, and Hf carbides (nitrides) in the near-surface regions may result in the embedding of the precipitates into the inwardly growing scale. Upon further scale thickening, these carbonitride particles

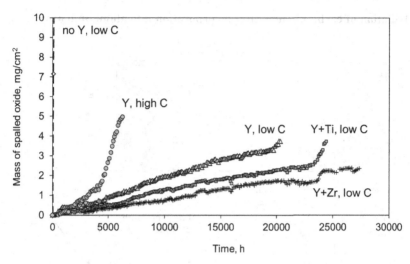

Fig. 10. Results showing effect of carbon and IVa group carbide formers on oxide scale spallation resistance in cyclic oxidation lifetime test (100 h cycles at 1200°C in laboratory air) of 1 mm thick coupons of model wrought Fe-20Cr-5Al-0.05Y base alloys. Tests were terminated upon macroscopic observation of breakaway oxidation following critical Al-depletion.

appeared to become oxidized within the scale (Fig. 11). This results in scale microcracking and formation of porosity, thereby leading to an enhanced scale growth rate (Fig. 12).[46]

3.5 *RE-Effects in Alumina Forming Coatings*

The RE-reservoir effects appeared to be not a unique feature of the FeCrAlY-alloys as similar observations were made with other alumina formers, in particular with MCrAlY (M = Ni,Co) coatings for gas-turbine applications.[44,47] Contrary to wrought FeCrAlY-alloys with typical RE-additions of 0.05%, the typical RE-contents in MCrAlY-coatings are 5–10 times higher in order to compensate for partial RE-oxidation during the various manufacturing steps. The latter include powder gas atomization, thermal spraying of the coating typically by low pressure plasma spray (LPPS) or high velocity oxy-fuel (HVOF) as well as post spray vacuum heat-treatment.[48] During all these manufacturing steps, the RE additions may (partly) oxidize. The variations of the RE-reservoir in the

Oxidation of carbo-nitrides after incorporation into aluminium oxide scale

$$t_1 \quad < \quad t_2$$

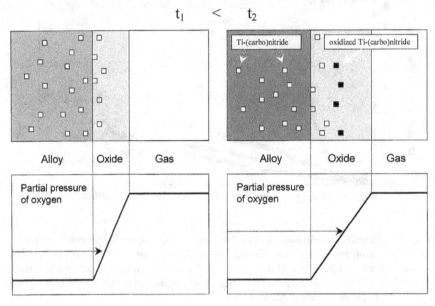

Fig. 11. Schematic of Ti carbonitride incorporation (t_1) and subsequent oxidation ($t_2 > t_1$) in inwardly growing alumina scale.[39]

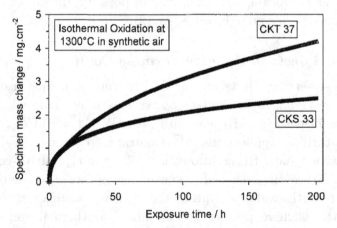

Fig. 12. Mass change data of two studied batches (CKS with low nitrogen content and CKT with high nitrogen content) of FeCrAl ODS alloy PM2000 obtained during isothermal oxidation in synthetic air at 1300°C.[46]

thermally sprayed MCrAlY coatings are difficult to control due to non-negligible tolerances of each processing step with respect to temperature, atmosphere composition (e.g., vacuum base pressure) as well as coating thickness.

Similar to the above mentioned RE-reservoir related thickness dependence of the scaling kinetics of FeCrAl-alloys (Section 3.3), the oxidation rates of MCrAlY-coatings were also found to increase with increasing coating thickness.[44] The latter effect was attributed to formation of bulky Y/Al-mixed oxides and internal yttria precipitates due to the higher RE reservoir in thicker coatings.

Another aspect of the RE-reservoir effects on oxidation of MCrAlY-coatings is the inhomogeneous formation of Y/Al mixed oxide pegs on rough surfaces produced by plasma spraying. Figure 13

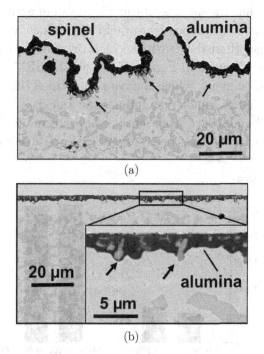

Fig. 13.   SEM image of free-standing, vacuum plasma sprayed NiCoCrAlY-coating containing 0.6%Y and 0.05%O: (a) as-sprayed surface and (b) ground surface after 25 h isothermal oxidation at 1100°C; arrows indicate Y-Al-mixed oxide pegs.[49]

shows that bulky Y-aluminate pegs are present almost exclusively in the concave regions of the rough coating surface. In contrast, the high surface to volume ratio and associated small local Y-reservoir in the convex coating regions resulted in a rapid Y-depletion leading to only minor amounts of very small Y-aluminate precipitates. An implication of the latter finding is that the oxide scale composition on rough MCrAlY-coating surfaces is per definition not uniform.

It could be shown[48] that the RE-reservoir can be substantially affected by the interaction of REs with oxygen introduced into the MCrAlY-coatings by the spraying process. The variations in the coating oxygen content may cause substantial differences in the cyclic oxidation lifetime of the ceramic thermal barrier coatings (TBCs) with MCrAlY-type bond coats. Figure 14 shows the lifetimes of the electron beam physical vapour deposited (EB-PVD) TBC systems. It is apparent that the lifetime of the TBC-system with a low oxygen content bond coat produced by spraying in "good" vacuum is substantially longer than that with the high oxygen content bond coat sprayed in "bad" vacuum. Figure 15 shows the SEM cross-sections of the failed specimens, which indicate that the TBC delamination occurred at a much smaller alumina scale thickness in the case of high oxygen bond coat. Apparently, the yttrium addition in the high oxygen bond coat has been largely tied up during the MCrAlY

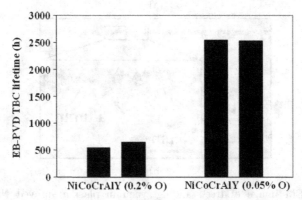

Fig. 14. Lifetimes of EB-PVD TBC systems on NiCoCrAlY bond coat with 0.3% yttrium and 0.05% and 0.2% oxygen respectively during cyclic oxidation test at 1000°C in air. Each bar represents an individual specimen.[48]

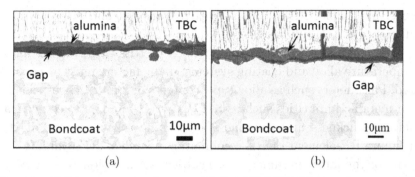

Fig. 15. SEM cross-section images of failed EB-PVD TBC systems on NiC-oCrAlY bond coat with 0.3% yttrium after cyclic oxidation at 1000°C in air: (a) bond coat with 0.2% oxygen after 600 h and (b) bond coat with 0.05% oxygen after 2500 h exposure.[48]

spraying process and consequently the alumina scale adherence is substantially poorer than that of the low oxygen bond coat.

It is noteworthy that in the case of TBC systems with the ceramic topcoat produced by air plasma spraying (APS) the higher oxygen content in the bond coat did not result in a lifetime shortening.[48] The reason for this is a different failure mechanism in APS-TBC's, whereby the crack propagation path leading to failure is partly located within the topcoat rather than at the TGO/bond coat interface.

4. Conclusions

The effect of REs on the scale formation and adherence has been extensively studied and exploited in practice. From the fundamental studies point of view, the change of the scale growth mechanism from mixed cation/anion to predominantly inward anion grain boundary transport as well as prevention of the adverse effect of sulfur and carbon impurities in degrading the oxide scale adherence are mostly supported explanations for the positive role of RE-additions. Even though the change in transport process and impurity gettering due to the presence of REs are accepted by many of the scientists dealing with the subject, there are still open questions on the details of the operating mechanisms. In particular, the exact way in which

RE-elements block the outward cation transport and improve scale adherence is not fully understood.

For successful practical applications of REs in advanced high-temperature alloy and coating systems, many factors must be considered. Parameters such as alloy type, contents of impurities (especially non-metallic impurities such as S, C, N, O), component geometrical factors (thickness and roughness) determine to a large extent the optimum RE-element or RE-combination as well as RE-contents. As many of the latter parameters are subject to unavoidable variations in large scale, multistep industrial processing, reproducibility of the optimized RE-additions represents a considerable technological challenge. The reproducibility issue is likely to become more important in the future as the alloys and coatings are likely to be used at higher temperatures than common today. Under such conditions, the oxide scale growth and spallation may become lifetime limiting factors in a large number of applications.

References

1. W.T. Griffiths and L.B. Pfeil, *Improvements in Heat Resistant Alloys*, U.K. Patent No. 459848 (1937).
2. W.J. Quadakkers and M.J. Bennett, Oxidation induced lifetime limits of thin walled, iron based, alumina forming, oxide dispersion strengthened alloy components, *Mater. Sci. Tech.* **10** (1994), pp. 126–131.
3. D.P. Whittle and J. Stringer, Improvements in high temperature oxidation resistance by additions of reactive elements or oxide dispersions, *Philos. Trans. R. Soc. London A* **295** (1980), pp. 309–329.
4. G.C. Wood and F.H. Stott, The development and growth of protective αFe-Al2O3 scales on alloys, in *Proc. Int. Conf. on High Temperature Corrosion NACE*-6 (2–6 March, 1981, San Diego, California, USA, 1983), pp. 227–250.
5. R. Prescott and M.J. Graham, The formation of aluminium oxide on high-temperature alloys, *Oxid. Met.* **38**(3/4) (1992), pp. 233–254.
6. H.J. Grabke, M. Siegers and V.K. Tolpygo, Oxidation of Fe-Cr-Al and Fe-Cr-Al-Y single crystals, *Zeitschrift für Naturforschung A* **50** (1995), pp. 217–227.
7. W.J. Quadakkers, A. Elschner, H. Holzbrecher, K. Schmidt, W. Speier and H. Nickel, Analysis of composition and growth mechanisms of

oxide scales on high temperature alloys by SNMS, SIMS and RBS, *Michrochim. Acta* **107** (1992), pp. 197–206.

8. W.J. Quadakkers, H. Holzbrecher, K.G. Briefs and H. Beske, Differences in Growth Mechanisms of Oxide Scales Formed on ODS and Conventional Wrought Alloys, *Oxid. Met.* **32**(1/2) (1989), pp. 67–88.

9. B.A. Pint, Experimental observations in support of the dynamic segregation theory to explain the reactive element effect, *Oxid. Met.* **45**(1/2) (1996), pp. 1–31.

10. J.G. Smeggil, Some comments on the role of yttrium in protective oxide scale adherence, *Mater. Sci. Eng.* **87** (1987), pp. 261–265.

11. P.Y. Hou, K. Prüßner, D.H. Fairbrother, J.G. Roberts and K.B. Alexander, Sulphur segregation to deposited Al2O3 film/alloy interface at 1000°C, *Scripta Mater.* **40**(2) (1999), pp. 241–247.

12. J.L. Smialek, *3rd International Conference on "Microscopy of Oxidation" Cambridge, September* 16–18th, 1996, *Proceedings*, S.B. Newcomb and J.A. Little (eds.) (The Institute of Materials, London, UK), p. 127.

13. J.D. Cawley, J.W. Halloran and A.R. Cooper, Oxygen tracer diffusion in single-crystal alumina, *J. Am. Ceram. Soc.* **74** (9) (1991), pp. 2086–2092.

14. C.R. Koripella and F.A. Kröger, Electrical conductivity of Al2O3: Fe+Y, *J. Am. Ceram. Soc.*, **69**(12) (1986), pp. 888–896.

15. J.L. Smialek and R. Gibala, Diffusion Process in Al2O3 scales: void growth, grain growth and scale growth, in *Proc. Int. Conf. on High Temperature Corrosion NACE*-6 (2–6 March, 1981, San Diego, California, USA, 1983), pp. 274–283.

16. J. Philibert and A.M. Huntz, Microstructural and diffusional studies in α-aluminas and growth mechanism of alumina scales in *Proc. 2nd Int. Conf. Microscopy of Oxidation*, (29–31 March 1993, Cambridge, UK), S.B. Newcomb and M.J. Bennett (eds.), (The Institute of Materials, London, 1993), pp. 288–297.

17. K.P.R. Reddy, J.L. Smialek and A.R. Cooper, 18O Tracer Studies of Al2O3 Scale Formation on NiCrAl Alloys, *Oxid. Met.* **17** (1982), pp. 429–449.

18. J. Jedlinski and G. Borchardt, On the oxidation mechanism of alumina formers, *Oxid. Met.* **36**(3/4) (1991), pp. 317–337.

19. R.A. Versaci, D. Clemens, W.J. Quadakkers and R. Hussey, Distribution and transport of yttrium in alumina scales on iron base ODS alloys, *Solid State Ionics*, **59** (1993), pp. 235–242.

20. A. Czyrska-Filemonowicz, R. Versaci, D. Clemens and W.J. Quadakkers, The Effect of Yttria Content on the Oxidation Resistance of ODS Alloys Studied by TEM; In *Proc. of 2nd Int. Conf. Microscopy*

of Oxidation (29–31 March 1993, Cambridge, UK) proceedings, S.B. Newcomb and M.J. Bennett (eds.) (The Institute of Materials, London, 1993), pp. 288–297.

21. B.A. Pint, A.J. Garratt-Reed and L.W. Hobbs, The reactive element effect in commercial ODS FeCrAl alloys, *Mater. High Temp.* **13**(1) (1995), pp. 3–16.

22. D. Naumenko, B. Gleeson, E. Wessel, L. Singheiser and W.J. Quadakkers, Correlation between the Microstructure, Growth Mechanism and Growth Kinetics of Alumina Scales on an FeCrAlY-Alloy, *Metall. Mater. Trans.* A **38** (2007), pp. 2974–2983.

23. D. Clemens, K. Bongartz, W.J. Quadakkers, H. Nickel, H. Holzbrecher and J.S. Becker, Determination of lattice and grain boundary diffusion coefficients in protective alumina scales on high temperature alloys using SEM, TEM and SIMS, *Fresenius J. Anal. Chem.* **353** (1995), pp. 267–270.

24. H. Nickel, W.J. Quadakkers, The Correlation between Growth Mechanisms and Technologically Relevant Protective Properties of Chromia and Alumina Scales on Oxide Dispersion Strengthened Alloys; *First Int. Conf. on Heat Resistant Materials*, (22–26 Sept., 1991, Wisconsin, USA, 1992), pp. 87–94.

25. J.P. Pfeifer, H. Holzbrecher, W.J. Quadakkers, W. Speier, Quantitative Analysis of Oxide Films on ODS-Alloys using MCs-SIMS and e-Beam SNMS, *Fresenius J. Anal. Chem.* **346** (1993), pp. 186–191.

26. C.W. Li and W.D. Kingery, Solute segregation at grain boundaries in polycrystalline Al2O3, in Structure and properties of MgO and Al2O3 ceramics, Advanced ceramics, *Am. Ceram. Soc.* **10** (1984), pp. 368–378.

27. W.J. Quadakkers, Growth mechanisms of oxide scales on ODS alloys in the temperature range 1000–1100°C, *Werkstoffe und Korrosion* **41** (1990), pp. 659–668.

28. H. Yoshida, Y. Ikuhara, T. Sakuma, Grain boundary electronic structure related to the high-temperature creep resistance in polycrystalline Al2O3, *Acta Mater.* **50** (2002), pp. 2955–2966.

29. D.J. Young, D. Naumenko, L. Niewolak, E. Wessel, L. Singheiser and W.J. Quadakkers, Oxidation kinetics of Y-doped FeCrAl-alloys in low and high pO2 gases, *Mater. Corros.* **61** (2010), pp. 838–844.

30. D.R. Siegler, Aluminium oxide adherence on Fe-Cr-Al alloys modified with Group IIIB, IVB, VB and VIB elements, *Oxid. Met.* **32**(5/6) (1989), pp. 337–355.

31. P.Y. Hou and J.L. Smialek, The effect of H2-anneal on the adhesion of Al2O3 scales on a Fe3Al-based alloy, *Microsc. Oxid.* **4** (2000), pp. 79–85.

32. H.J. Grabke, D. Wiemer and H. Viefhaus, Segregation of sulphur during growth of oxide scales, *J. Appl. Surf. Sci.* **47**, pp. 243–250 (1991).
33. E. Schumann, J.C. Yang and M.J. Graham, Direct observation of the interaction of yttrium and sulphur in oxidised NiAl, *Scripta Mater.* **34**(9) (1996), pp. 1365–1370.
34. A.B. Anderson, S.P. Mehandru and J. Smialek, *J. Electrochem. Soc.* **132** (1985), pp. 1695–1708.
35. W.J. Quadakkers, H. Holzbrecher, K.G. Briefs and H. Beske The Effect of Yttria Dispersions on the Growth Mechanisms and Morphology of Chromia and Alumina Scales Europ. Coll. on the Role of Active Elements in the Oxidation Behaviour of High Temperature Alloys, 12–13 Dec. 88, Petten, NL Proc., European Communities, Edited by E. Lang, pp. 155–173.
36. H.J. Grabke in *Proc. Int. Conf. on Metal Supported Automotive Catalytic Converters*, Wuppertal, Germany, 1997, H.Bode (ed.) (Werkstoff-Informationsgesellschaft mbH, Frankfurt, 1997), p. 139.
37. J. Sringer, The reactive element effect in high-temperature corrosion, *Mater. Sci. Engin. A* **120** (1989), p. 129.
38. W.J. Quadakkers, J.F. Norton, H.J. Penkalla U. Breuer, A. Gil, T. Rieck and M. Hänsel, SNMS and TEM-Studies Concerning the Oxidation of Cr-Based ODS Alloys in SOFC Relevant Environments, *3rd Int. Conf. on "Microscopy of Oxidation"* Cambridge, September 16–18th, 1996, *Proc.*, S.B. Newcomb and J.A. Little (eds.) (The Institute of Materials, 1996), pp. 221–230.
39. W.J. Quadakkers and L. Singheiser, Practical Aspects of the Reactive Element Effect, *Mater. Sci. Forum* **369–372** (2001), pp. 77–92.
40. D. Naumenko, V. Kochubey, L. Niewolak, A. Dymiati, J. Mayer, L. Singheiser and W.J. Quadakkers, Modification of alumina scale formation on FeCrAlY alloys by minor additions of group IVa elements, *J. Mater. Sci.* **43** (2008), pp. 4550–4560.
41. J. Klöwer, Factors affecting the oxidation behaviour of thin Fe-Cr-Al foils Part II: The effect of alloying elements: Overdoping, *Mater. Corros.* **51** (2000), pp. 373–385.
42. B.A. Pint, Optimization of Reactive-Element Additions to Improve Oxidation Performance of Alumina Forming Alloys, *J. Am. Ceram. Soc.* **86**(4) (2003), pp. 686–695.
43. E. Wessel, V. Kochubey, D. Naumenko L. Niewolak, L. Singheiser and W.J. Quadakkers, *Scripta Mater.* **51**(10) (2004), pp. 987–992.
44. J. Toscano, R. Vaßen, A. Gil, M. Subanovic, D. Naumenko, L. Singheiser and W.J. Quadakkers, Parameters affecting TGO growth and adherence on MCrAlY-Bond coats for TBC's, *Surf. Coat. Technol.* **201** (2006), pp. 3906–3910.

45. V. Kochubey, D. Naumenko, E. Wessel, J. Le Coze, L. Singheiser, H. Al-Badairy, G.J. Tatlock and W.J. Quadakkers, Evidence for Cr-carbide formation at the scale/metal interface during oxidation of FeCrAl-alloys, *Mater. Lett.* **60** (2006), pp. 1654–1658.

46. W.J. Quadakkers, D. Naumenko, L. Singheiser, H.J. Penkalla, A.K. Tyagi and A. Czyrska-Filemonowicz, Batch to batch variations in the oxidation behaviour of alumina forming Fe-based alloys, *Mater. Corros.* **51** (2000), pp. 350–357.

47. T.J. Nijdam and W.G. Sloof, Effect of Y distribution on the oxidation kinetics of NiCoCrAlY bond coat alloys, *Oxid. Met.* **69** (2008), pp. 1–12.

48. P. Song, D. Naumenko, R. Vassen, L. Singheiser and W.J. Quadakkers, Effect of oxygen content in NiCoCrAlY bond coat on the lifetimes of EB-PVD and APS thermal barrier coatings, *Surf. Coat. Technol.* **221** (2013), pp. 207–213.

49. A. Gil, V. Shemet, R. Vassen, M. Subanovic, J. Toscano, D. Naumenko, L. Singheiser and W.J. Quadakkers, The effect of surface condition on the oxidation behaviour of MCrAlY coatings, *Surf. Coat. Technol.* **201** (2006), pp. 3824–3828.

Index

A

2024-T4 alloy, 130
21/4Cr-1Mo alloy, 29, 121
214Cr-1Mo alloy, 125
800HT alloy, 94
9Cr-1Mo alloy, 29, 121
253MA, 89, 91, 97
353MA, 89, 92, 94
9Cr-ODS alloy, 79, 82
1st generation alloy, 140
4th generation alloy, 141
5th generation alloy, 142
α-alumina, 15
absorber coatings, 194
acceptors, 247
accidents, 27
active element effect, 26
active elements, 7
adequate strength, 27
adherence, 256
advanced analytical tools, 201
aerospace gas turbines, 23
aerospace materials, 201
aggressive environment, 94
air plasma spraying, 263
aircraft corrosion, 130
aircrafts, 129
airframes, 134
AISI 304H, 67

AISI Type 304, 13
Alloy 904, 39
alloy development, 134
alloying design, 58, 84
alloying element, 10
alternate energy source, 192
alternating stresses, 149
alumina, 11, 28, 80, 114, 174
alumina films, 170
alumina former, 146
alumina layer, 14
alumina single-crystals, 247
aluminide coatings, 172, 174
aluminum oxide, 97
aluminized coatings, 186
aluminizing, 187
analytical techniques, 246
anchoring, 246
annealing, 89
anthracite coal, 107
anti-reflection coating, 195
antiphase boundary, 75
Arrhenius equation, 41
ash, 20
ash deposit, 20, 121
ash erosion, 114
ash formation, 106
atomic radii, 74
austenitic materials, 95

Index

sulfates, 21

sulfate-type, 120

sulfidation, 1, 15, 20, 29–30, 89, 105, 111, 121

sulfidation resistance, 29

sulfide, 110

sulfur, 95, 104–105, 109, 120, 124

sulfur activity, 20

sulfur dioxide, 15, 18, 21, 49

sulfur effect, 26–27

sulfur segregation, 252

sulfur trioxide, 21

sulfur vapor, 15

super critical, 125

super heaters, 110, 120–121, 125

superalloys, 11, 28, 35–36, 135, 162

supercritical boilers, 108

superheated tubes, 21

superheated vapor, 108

superheater, 9, 112, 114

superheater tubes, 20, 38, 68

superheater/reheater, 120

superheaters, 121

superpartial dislocations, 75

surface modification, 231

surface modification techniques, 168

surface oxidation, 92

sustainability, 30

synergistic effect, 54

synthetic gas, 89

T

T91, 64–65

T92, 64–65

TBC systems, 262

temperature gradient, 137

tempering, 60

tensile, 33

tensile strength, 35

tertiary creep, 41

TGO, 176, 178

thermal barrier coating (TBC), 152, 175

thermal cycling, 6–7, 26

thermal efficiency, 57

thermal expansion, 97, 122

thermal gradient coatings, 156

thermal mechanical fatigue, 149

thermal power, 101, 161, 171

thermal spray, 121, 168

thermal spray coating, 29

thermal stability, 221

thermal stresses, 25–26

thermally activated, 41

thermally grown oxide, 176

thermally grown oxide layer, 154

thermally induced stresses, 136

thermodynamic stabilization, 222

thermomechanical processing, 76

thickening rate, 252

third generation, 141

third-element effect, 232

TiN coatings, 170

titanium, 9

TMS-75, 72

total positive ion image, 206

tracer experiments, 246

transient oxidation, 10

transportation, 161

transverse grain boundaries, 71

trisulfate attack, 110

trisulfates, 111–112

tritium permeation, 184

tube failures, 114

tungsten coating, 180

turbine, 102

turbine blades, 99, 102, 147

turbine discs, 204, 206

turbine disks, 149

turbine efficiency, 132

Printed in the United States
By Bookmasters